AI POWERED

LEADERSHIP

for

MODERN FAMILIES

By

Dr. Wesley Carter

Copyright Page

© 2025 Dr. Wesley Carter

All rights reserved.

ISBN: 979-8-218-78235-1

Printed in the United States of America

For information about permissions, workshops, or speaking
engagements, visit:

www.DrWesleyCarter.com | www.KidsByCarter.com

Testimonials

"Before discovering Dr. Wesley's approach, I thought technology was something I needed to protect my children from. Now I see it as something I can use to lead them with. Learning how to integrate AI and digital tools into my parenting completely changed our household dynamic. Instead of reacting to every challenge, I've learned to pause, reflect, and use technology to help my kids build emotional awareness, structure their routines, and communicate more effectively.

This isn't about screen time—it's about smart time. It's about modeling balance, curiosity, and confidence in a world driven by data. Dr. Wesley's framework helped me see that parenting with technology isn't a shortcut; it's a strategy. I finally feel like I'm leading my family, not just managing it."

—S. Pamela Gray, Retired Family Court Judge

"I've been using AI for years in my work. But, Dr. Wesley taught me how to utilize technology with my grandchildren. Her ABCD Framework helped me create conversation prompts to discover what matters most to my grandchildren. What I appreciate most is how Dr. Carter reframed AI from a tech tool into a relationship tool. She helped me use AI prompting to demonstrate empathy, curiosity, and presence. Now, our talks aren't just about technology; they're about life, purpose, and legacy."

—Gary Moore

"Dr. Wesley's book opened my eyes to what it really means to be an intentional parent in a digital world. I used to feel guilty about how much technology surrounded our family, but now I see it as an opportunity to teach leadership, problem-solving, and emotional intelligence right at home."

—*Alexis Bartlett*

Dedication

To my beautiful, intelligent, charming, witty, beloved children:

Thank you for giving me every possible challenge, curiosity, and joy that parenting could offer and for keeping our story filled with lessons, laughter, and unforgettable moments. You have been my greatest teachers, my mirrors, and my proof that love can stretch farther than reason, rest, or routine ever could.

You were my living classroom, testing every principle, expanding every ounce of patience, and reminding me that love grows strongest where accountability and grace meet. Every tear, triumph, and late-night prayer drew me closer to God, who guided me through the moments I wasn't sure how to navigate. You taught me true leadership is sometimes unpopular, but always purposeful.

Every insight in this book was learned with you, sometimes joyfully, sometimes the hard way, but always with purpose. You made me wiser, steadier, and more intentional. You turned ordinary days into masterclasses in humility, humor, and heart. And though I was never the permissive parent, I was the loving one: the one who believed in your potential too much to let comfort replace character, and who loved you enough to say "no" when it mattered most.

And now, as you move confidently into your own lives, I smile with pride. Your independence and boldness are the ultimate signs I did my work well, I built a foundation of faith, discipline, and resilience strong enough for you to stand on. I thank God you can dream freely, work diligently, and love deeply, knowing every boundary I held and every expectation I set came from love, not leniency.

May you never forget that your comfort was built from sacrifice and grace was polished by time and truth. Because of you, I learned true legacy is grounded in perseverance, laughter, and love; and it outlasts misunderstanding. When your own children ask what love looks like, show them it leads firmly, forgives freely, and keeps choosing family again and again.

Because of you, I didn't just write about parenting, I *earned* it.

With faith, gratitude, laughter, and love beyond measure,

Mommy & Grandmommy

✳✳✳

To the ones I have had the privilege to mother and grandmother along the way:

Along the way, I also *mothered* and *grandmothered* others: beautiful souls who entered my life through purpose rather than lineage. You came to me as students, mentees, neighbors, friends, or simply wanderers looking for light, and somehow, in the sacred rhythm of time, we became family.

Many of the stories, reflections, and lessons woven into these pages grew from our shared moments: our talks that stretched late into the night, our laughter through trials, our tears in seasons of growth. Each experience with you has shaped not only my understanding of parenting, but also of humanity, faith, and grace.

Though we are not bound by blood, we are connected by choice, love, and divine design. We have forged *fictive kinships* that are every bit as real and powerful as the ties of ancestry. You have allowed me to see the ripple of motherhood extend far beyond biology: into mentorship, compassion, and legacy. You, too, are my daughters and sons.

And I love you fiercely.

Mommy & Mimi

✳✳✳

To my fellow parents:

This book is also dedicated to the countless parents who have opened their hearts, homes, and stories to me. Whether through my coaching sessions, training programs, or speaking engagements, you have allowed me to witness the raw, remarkable truth of parenthood: the victories, the uncertainties, the humor, and the holy work of simply showing up each day.

Your honesty, vulnerability, and courage have transformed lessons into wisdom. Through your stories, I have learned leadership at home is not about control but about connection; not about getting it perfect, but about being purposeful. You've shown me every family, regardless of size, shape, or structure, holds infinite potential for love, learning, and leadership.

Some of the stories and experiences within these pages were born from our shared moments: from the questions you've asked during workshops, the breakthroughs you've shared during consultations, and the reflections you've offered in those sacred, quiet spaces between learning and living. Your openness has turned professional exchanges into personal revelations, and for that, I am forever grateful.

Together, we are redefining what it means to lead our families with grace and grit. We are exploring new territory: learning to use artificial intelligence not as a replacement for human touch, but as an ally for insight, empathy, and confidence in an ever-changing world.

Thank you for walking this journey with me, for trusting me to speak into your parenting, and for teaching me just as much in return. You are proof that community, courage, and curiosity can raise not only strong children, but stronger leaders in us all.

With deep respect and shared purpose.

Dr. Wesley

Foreword

As a family therapist, I have spent over a decade sitting across from parents, grandparents, educators, and children, all searching for the same thing: connection. In an age where technology often feels like it is pulling families apart, AI Powered Leadership for Modern Families arrives as a bridge. It shows us how to use technology not as a distraction, but as a tool for emotional clarity, meaningful communication, and intentional leadership at home.

Dr. Wesley Carter has written more than a parenting book. She has created a leadership manual for modern family life. Every page redefines what it means to guide, nurture, and influence the next generation in a world that moves at the speed of data. Her unique approach integrates emotional intelligence, modern leadership strategies, and the thoughtful use of Artificial Intelligence to strengthen family bonds rather than weaken them.

What makes this work remarkable is its accessibility. Whether you are a school counselor helping students manage social stress, a social worker supporting families in crisis, or a district leader shaping programs for student well-being, this book belongs in your hands. It gives professionals and caregivers a shared language for empowerment, accountability, and empathy.

Parents and grandparents will find in these chapters both comfort and challenge. Comfort in knowing that the principles of leadership they use at work can also guide their family relationships. Challenge in realizing that emotional growth and digital literacy must coexist if we want to raise resilient, self-aware young people.

AI Powered Leadership for Modern Families deserves a place in every school library, counseling office, and community center. It is a roadmap for those who believe family leadership is not about control but about cultivating wisdom, curiosity, and connection in the age of Artificial Intelligence.

Dr. Carter reminds us that leadership begins at home and that with the right mindset and tools, every parent, teacher, and grandparent can shape a future defined not by fear of technology but by the strength of human love and intentional guidance.

Erica Allen, MFT, LPC-A

Dallas, Texas

Preface

This book was born from a question: *What would happen if parents started leading their families the same way effective leaders guide organizations, with vision, empathy, and intention?*

I see an extraordinary opportunity to improve modern parenting through the thoughtful use of artificial intelligence. But not without structure and purpose. That is why I created the ABCD Framework, a scalable and memorable model that helps parents use AI with clarity, empathy, and intention. It transforms random interaction into reflective guidance, turning every prompt into a tool for growth. With the ABCD Framework, AI becomes more than technology; it becomes a tool for emotional intelligence and leadership at home.

As a mother, technologist, and leadership coach, I wrote this book because I believe family life deserves the same strategic attention we give to our careers. Too often, parents reserve their strongest skills such as planning, communication, emotional intelligence, and problem-solving for the workplace, while home becomes the place we simply react and recover. Yet the same abilities that make us high performers at work can transform the way we lead our families. Through the ABCD Framework, parents can create purposeful AI prompts that help children grow academically, intellectually, emotionally, and socially.

When we apply our leadership mindset at home, we turn family life into a place of clarity, compassion, and continuous growth.

This book is about using technology and leadership together to elevate modern parenting. Most, if not all, professionals rely on technology to excel in their work, so why not bring that same advantage home? My hope is that these pages help you see yourself as the leader of a living, breathing organization called family and inspire you to lead it with grace, humor, and heart. When used with purpose, technology can help you guide your home with the same clarity, confidence, and creativity that drive your professional success.

Welcome to the next chapter of family leadership. Let's grow smarter together.

Table of Contents

PART I
SMARTER PARENTING

PART II
LEADING THROUGH THE REAL-WORLD CHALLENGES

PART III
LIVING LEADERSHIP

INTRODUCTION

Kids by Carter: AI-Powered Leadership for Modern Families

Parenting today is not a passive hobby; it is the ultimate startup venture. Every morning begins with strategy meetings over breakfast and ends with performance reviews at bedtime. Between homework battles, emotional negotiations, and the magnetic pull of social media, raising children now requires the patience of a saint, the agility of a founder, the insight of a psychologist, and the strategic vision of a Fortune 500 CEO. The modern family leader must manage competing priorities, shifting moods, and ever changing algorithms while holding the heart of the household together.

Artificial Intelligence (AI) has entered this landscape as both a challenge and a gift. The challenge lies in keeping up with the speed of technology and its influence on our children's minds. The gift is the opportunity it gives us to lead with greater awareness, foresight, and calm. When used wisely, AI becomes more than a tech tool; it becomes a mirror, a mentor, and a resource that helps parents and grandparents think clearly before reacting, design better systems for family life, and model thoughtful leadership in a digital world.

I wrote this book because parents and grandparents are not just caregivers; we are leaders, visionaries guiding a living, breathing organization called family. Like effective executives, we need systems, clarity, and tools that reflect our mission and values. Our homes may not have boardrooms, but they do have cultures, goals, and team dynamics that shape how every member thrives.

Every parent and grandparent can become an intentional, informed leader who raises children capable of navigating complexity with character and confidence. Through the principles and AI prompts in this book, you will discover how to strengthen emotional intelligence, encourage independent thinking, and use technology as a force for growth rather than distraction. Together, we will redefine what leadership means at home.

I wrote this because I do not have all the answers. I wrote it because I have made almost every mistake in the book. My four children, now a physician, an attorney, a program manager, and a gamer, were my original test group, and let's just say a few learning opportunities came cleverly disguised as disasters. Some of my parenting pilots are now cautionary tales. What I learned along the way is great parents are rarely perfect; they are present, as much as possible. And timeouts are sometimes for the parents, not the kids. Leaders learn, adjust, and keep leading, even under duress.

Now, I am the parent of adults, and in many ways it feels like I am beginning again. The landscape has shifted, and I am learning new ways to connect with the people I once guided through every decision. At times I still feel unprepared, but I am also aware that this stage is an invitation to evolve, to lead with more listening, gentler guidance, and greater respect for their independence. The rules may be new, yet the goal remains the same: to nurture relationships built on trust, empathy, and mutual growth. This season of parenting reminds me that leadership is never finished; it simply changes form, and I continue to rely on my ABCD Framework because it remains both relevant and required to strengthen communication every step of the way.

When I launched my Kids by Carter consulting business in 2010, I started writing newspaper columns about the real trials and triumphs of raising children, whether alone, with a partner, or with the help of a village. Those articles blended my own lived experience with peer-reviewed research about child development and the critical role parents, grandparents, and guardians play in making intentional, impactful decisions for their families.

Much of what you will read in this book is drawn from years of learning, achieving, failing, reflecting, and sharing. Some stories come directly from my own parenting journey, while others come from the experiences and wisdom of fellow parents and grandparents who opened their hearts to me. I am grateful for their honesty, because it was in those shared challenges we found our greatest strength and community.

My qualifications? Let's just say I have worn nearly every hat available: child, parent, grandparent, godparent, educator, consultant, professor, technologist, and chaos coordinator. I have been the teacher, the coach, the boss, the student, the manager, and the leader making sure safety nets were deployed when it mattered most. Another qualification is that I wanted to be here, at this moment in time, introducing the thoughtful use of AI in parenting. I was eager to do the research and the work needed to provide a balanced, human-centered view of how technology can strengthen family leadership rather than replace it.

As a technologist, leadership consultant, and founder of Kids by Carter, I have worked with parents, grandparents, educators, and administrators to build partnerships that helped families thrive. My journey into this work began when my church invited me to speak to parents about how I was supporting the success of my high achieving children. I was simply leading my children in a way that reflected intention, structure, and love, and it was noticeable. That moment opened the door to a lifelong mission. Since then, I have taught workshops on advocacy, relationship building, navigating school systems, and strengthening home school collaboration, all grounded in one belief: parents should be the first and most intentional leaders in a child's life.

Academically, I hold an MBA and a doctorate in management with a specialization in leadership, degrees I earned while working full time, surviving a divorce, raising my children on three to four hours of sleep a night, driving carpools, volunteering, and somehow maintaining a 3.86 GPA. Not to brag, but to show that I understand what it means to lead under pressure. In addition to my degrees, I have earned more than ten certifications in technology and leadership, because I believe learning never stops. I am a lifelong student of both people and progress. The same leadership principles that carried me through those years can help you guide your own family with clarity, confidence, and grace, powered by both real and artificial intelligence.

Professionally, I am a technologist, someone who designs, manages, and applies technology to solve human problems. I drove innovation at Fortune 100 companies such as IBM and Cisco, leading projects where data and design influence human behavior. I did not just watch AI change the world; I led the teams that power it. So when I tell you that AI affects how we pay attention and make choices, including why it can be so hard for your child to put the phone down, I am speaking from the engine room, not the lobby.

Personally, I have logged thousands of hours using AI tools at home: organizing bills, managing healthcare, planning travel, and even coaching myself through tough conversations. I have used AI to help me write letters I was too emotional to start, prepare for important meetings, and offer support to my friends, family, and clients. I have seen how these tools can bring order, offer wise counsel, reduce stress, and free up time for what really matters. But I also know how intimidating it can feel to begin. My goal is to make the use of technology approachable, practical, and flexible to help you harness AI confidently as a support tool for family leadership.

Every day, AI helps me practice emotional intelligence in real time. I use AI prompts to help me reflect on my tone before sending a message, to reframe frustration into understanding, and to explore language that turns conflict into connection. When I am tired or reactive, I ask AI to help me find words that sound like the calm, thoughtful leader I want to be. These small moments of digital coaching help me notice my triggers, name my emotions, and model self-regulation. Over time, AI has become a tool and a mirror, reminding me leadership begins with emotional awareness.

Through my ABCD Framework, I have learned how to bring both technology and heart into harmony. By combining the analytical power of AI with the emotional depth of human leadership, I created a way to keep compassion and clarity at the center of every digital interaction. What I have learned is a gift I now offer to parents, grandparents, and guardians who are leading their own families through the same challenges and joys.

The ABCD Framework is my gift to you, offered with the hope that it will help you avoid some of the mistakes I have made and discover a more thoughtful, grounded way to lead. It is how I use AI prompts to build connection, embody empathy, and guide family decisions with leadership and love. I am going to show you how to do the same, so you can strengthen your family's intellect, emotional intelligence, communication, and sense of purpose.

What This Book Will Teach You

This isn't a technology manual; it's a leadership playbook for your family. You won't find fearmongering about screens or lectures about "the good old days." Instead, you'll learn how to use AI as a tool for empathy, accountability, and problem-solving, the timeless qualities every great leader and parent needs.

Parenting is leadership in its most personal and powerful form. Every decision you make, every boundary you set, and every conversation you begin creates direction for your family. Each of these moments becomes a lesson in how to respond to life with awareness, intention, and care. This book invites you to see yourself as the chief architect of your child's growth. Leadership at home is not about control. It is about creating the conditions where curiosity, confidence, and resilience can flourish.

The AI prompts in this book are designed to help you and your children grow together. You can use the same framework to prepare for parent teacher meetings, handle difficult conversations, or reflect on your own leadership. Each time you use the four steps of the ABCD Framework, you are practicing intentional thinking, the kind that helps you move from reaction to reflection and from correction to connection.

Throughout this journey, I will refer to you as a parent leader because there is an important difference. A parent responds to what is happening. A parent leader guides what happens next. My goal is to help you lead your family with confidence, curiosity, and calm. You are doing more than managing your household. You are shaping the next generation of thinkers, problem solvers, and emotionally intelligent human beings. That kind of leadership requires courage, intellect, and heart.

Many parents will invest in meals, vacations, and the newest technology but hesitate to invest in a resource that can transform their children's direction. For less than the cost of one meal, this book gives you access to more than two hundred AI prompts written in a clear, practical ABCD Format. These prompts will help you lead your family into a new season of understanding, growth, and peace. You are already thinking differently, and that mindset sets you apart as a parent leader.

This book blends my lived experience with academic training, consulting expertise, and years of driving innovation and practicing leadership. It gives you practical tools to raise thoughtful, emotionally aware, successful, and socially responsible young people who are ready to thrive in a digital world.

Together we will explore how AI can help you create structure that supports creativity, make confident decisions, and communicate with compassion even when emotions are strong. You will learn how to build a household that functions like a thriving organization where every member has a voice, a role, and a purpose. As the leader of your family, you will see how to inspire collaboration, encourage accountability, and develop self-leadership in your children.

You will also learn how to personalize your family leadership framework using simple, human centered systems supported by

AI. AI prompts embedded in the ABCD Framework will help you create daily habits that strengthen connection, curiosity, and confidence. As you blend wisdom, warmth, and technology, you will build a family culture grounded in empathy, accountability, and love, the kind of leadership that nurtures children who are empowered to improve the world.

The ABCD Framework for Smarter Parent Leadership

Even the most patient parent runs out of words at ten o'clock at night when everyone is tired, the Wi-Fi is slow, and the dishes are still in the sink. Someone cannot find a charger, a sibling argument is escalating, and bedtime negotiations start to sound like a peace summit. That is when artificial intelligence becomes more than a convenience. It becomes a coaching tool for emotional clarity. It helps you think before you speak, breathe before you react, and model calm leadership when everyone else is falling apart.

Research shows that parents who use AI prompts with intention strengthen not only their decision-making skills but also their emotional intelligence. Studies reveal that AI-guided reflection helps both parents and children practice empathy, manage emotions, and communicate more effectively. When you begin to use AI prompts in the structured ABCD Framework to explore feelings, choose language carefully, or plan difficult conversations, they model the kind of emotional awareness children need to thrive. AI does not replace intuition. It enhances it by creating space to pause, reflect, and respond with wisdom.

AI becomes your point of reference, a tool that helps you pause before reacting, reframe before repeating, and plan before problems pile up. Used with care, it shows your children that true leadership begins with reflection. Effective leaders use strategic tools to clarify their vision, make thoughtful decisions, and ensure that their actions reflect their values. This is a skill every child should see in practice, and it begins with you.

I created a tool I want to share with you as a parent leader. It is practical, memorable, and grounded in leadership principles. It is called the ABCD Framework, and it transforms how parents guide their children's growth and development. The ABCD Framework gives parents a simple and repeatable way to lead their families with clarity and confidence. It is structured as an easy-to-remember process that you can use even in moments of stress when emotions are high and quick decisions matter. Each step represents a way to think and act like an effective leader who nurtures intellectual curiosity, strengthens emotional intelligence, and supports healthy growth.

The Four-Part ABCD Framework Prompt Structure

When you need guidance that feels personal and practical, begin with the ABCD Framework. Here's the secret sauce: when you use all four parts of the **ABCD Framework**, AI stops acting like a trivia bot and starts acting like a thoughtful reference tool built just for you.

Actor: Tell AI who you want it to act as so that you receive the kind of perspective you need. For example, say "You are a Family Therapist" or "You are a Leadership Coach for parents." This step sets the point of view that shapes the tone and direction of the guidance.

Background: Describe your situation clearly. Explain what is happening, who is involved, and how you are feeling. Give AI the same amount of detail you would share with a trusted professional. The more complete the context, the more relevant and useful the response will be.

Call to Action: Step into your leadership role and give AI a clear direction. Decide what you need most such as examples, insights, scripts, conversation starters, or specific solutions. Be clear so the response aligns with your goal. Leadership begins with clarity. Start your Call to Action with verbs such as prepare, provide, design, create, teach, build, guide, encourage, inspire, strengthen, and reflect. Do not limit yourself to these words. As a parent leader, choose the verb that matches your family's needs and the kind of growth you want to create.

Data Check: Strong leaders trust but verify. Before you press enter, invite AI to ask questions before giving advice. This step keeps the feedback focused on your family rather than a generic response. Always close your prompt with the statement, "If you need more information, ask me questions before proceeding." This simple practice teaches accuracy, thoughtfulness, and curiosity. These are the same qualities that form the foundation of emotional intelligence and effective leadership.

When you ask AI to request more details before giving answers, you are modeling the same kind of critical thinking and curiosity you want your children to practice. You are showing them that wisdom begins with good questions, not quick conclusions.

Tip: AI will never know your child better than you do. It cannot understand your family's humor, history, or heart. What it can do is organize information, conduct research, and help you see possibilities you may not have considered. Use AI to broaden your thinking and strengthen your choices. Try creating a prompt before a difficult conversation or family meeting. You will approach the moment calmer, clearer, and more confident.

Over time, using the ABCD Framework will change how your family communicates. Your children will stop seeing you as the person who has all the answers and start seeing you as a calm, curious guide who leads with empathy and structure. That kind of leadership does not simply solve problems. It builds people.

Five Principles of Family Leadership

Every organization has a leadership philosophy, and so should your family. The five principles of family leadership outlined here form the foundation of this book. They shape every story, strategy, and AI prompt you will encounter. Each principle invites you to lead with empathy, consistency, and courage in the digital age. Together, they provide a compass for raising children who are intellectually curious, emotionally aware, and socially responsible in a world shaped by rapid change and advancing technology.

1. **Emotional Intelligence Before Action: The Core of Composed Leadership**

   Setting the emotional tone at home is one of the most important acts of leadership. Just as strong leaders shape the culture of their organizations, parent leaders shape the emotional climate of their families through their self-awareness, empathy, and communication. Calm, predictable reactions build safety and trust, while emotional control teaches children strength and gentleness can coexist. When you lead with composure, compassion, and accountability, you model emotional maturity. Every family ritual, shared value, and positive tone becomes an anchor of stability. Emotional tone is never accidental. It is the intentional result of a parent leader who manages energy, leads with awareness, and treats every interaction as an opportunity to model the culture they want their children to inherit.

2. **Clarity is Kindness: Building Safety Through Structure**

Clarity is a powerful expression of leadership and emotional intelligence. When parents communicate expectations with understanding, they prevent confusion and fear. Strong leaders give direction with empathy, explaining what matters and why it matters. Boundaries protect relationships when they are established with peace and purpose. When expectations are clear, consistent, and reinforced with compassion, children learn responsibility and self-discipline. Inviting them to help define family rules builds ownership and mutual respect. Pairing structure with empathy teaches that clarity is about care. Clarity is kindness in action because it replaces uncertainty with safety, turns assumptions into understanding, and allows every family member to know where they stand and why it matters.

3. **Process over Perfection: Form is the Foundation of Mastery**

Leadership at home is not measured by flawless outcomes but by how thoughtfully we approach each challenge. The process is where mastery begins. When parents focus on form, on how they listen, question, and respond, they teach that steady effort matters more than instant success. Modeling problem solving turns daily challenges into lessons in resilience and reflection. Rather than rushing to fix every issue, wise parents invite dialogue, guide exploration, and help children reason through their choices. This process transforms mistakes into opportunities for growth. Technology can strengthen this practice by helping families gather information, explore possibilities, and view challenges from multiple perspectives. When parents talk through decisions out loud, children witness emotional intelligence in action. They learn that progress takes patience and that mastery lives in the process itself, the rhythm of trying, learning, and trying again with greater understanding.

4. **Empathetic Accountability: The Balance of Heart and Hold**

Empathetic accountability means leading with both understanding and expectation. It is the balance of heart and hold, where love guides responsibility and compassion strengthens discipline. True leadership requires parents to see beyond behavior to the emotions beneath it while still maintaining boundaries that shape growth. J. Bert Freeman, author of *Taking Charge of Your Positive Direction*, teaches positive direction is about guiding people toward growth rather than focusing on mistakes. When applied to parenting, this principle helps children experience accountability as support rather than discouragement. When your child makes a poor choice, invite reflection instead of reaction by asking, "What did you learn from this?" or "How can we handle this better next time?" These forward focused questions shift the conversation from judgment to growth. By pairing empathy with structure, you turn correction into coaching and responsibility into self respect. This approach teaches that accountability is not control but care in motion, the kind of leadership that builds trust, integrity, and emotional maturity for life.

5. **The Growth Mindset: Leadership Lives in Iteration**

A growth mindset reflects the heart of leadership because true progress happens through steady iteration. Parents who lead with this mindset understand that growth is a continual process of learning, adjusting, and trying again. Every conversation, challenge, and moment of correction becomes a chance to practice patience, resilience, and reflection. When children say, "I can't," an emotionally intelligent leader helps them add the word "yet." By praising persistence instead of perfection, curiosity instead of quick answers, and courage instead of comfort, you show that leadership is built one effort at a time. Growth minded families embrace the process, knowing that mastery is not a single achievement but the result of consistent practice guided by love and learning. This is how parent leaders cultivate confidence, turning small steps into lasting success and everyday effort into the foundation of lifelong growth. This is how leaders raise leaders, teaching them how to turn frustration into fuel and mistakes into motivation.

At the center of family leadership is a simple truth: children learn who to become by watching how we lead. When you set the emotional tone with consistency, provide clarity with kindness, model thoughtful problem solving, hold accountability with empathy, and cultivate a growth mindset, you are doing more than managing your home. You are shaping a legacy. Every moment becomes a lesson in emotional intelligence, responsibility, and resilience. You show love has structure, discipline has compassion, and growth is a continuous process. In the end, great parenting is not about perfection. It is about direction. You are raising leaders who know how to think, feel, and act with purpose, even when no one is watching.

A Final Word Before We Begin

I will not promise you will become a flawless parent, because I am not one either. I am still learning, still making mistakes, and still practicing how to be a better leader for the people I love. You will mess up. You will second-guess yourself. You will say things you wish you had phrased differently, forget events you meant to remember, and sometimes wonder if your best will ever be enough. But with the tools in this book, every misstep will move you forward. You will recover faster, reflect deeper, and lead with greater love and clarity.

I often think about how different my parenting journey might have been if AI had been available when I was actively raising my children. I cannot go back, however, **you** can move forward. You can lead your family with the advantage of insight, structure, and support technology now makes possible. AI prompts do not remove the hard parts, but they make the process more intentional and less isolating.

Parent leadership is the hardest leadership role any of us will ever hold, and the only one where our legacy learns to walk, talk, and think for itself. There are no performance reviews, no bonuses, and often no immediate feedback. Sometimes the results do not show for years. Yet every day, you show up. You manage emotions, motivate your team, and make decisions to shape the character and confidence of the people you love most. It is leadership without an operator's manual, accountability without applause, and progress measured not in profits but in people.

Technology, when used with purpose, becomes your tool. It can help you track goals, plan with intention, and model calm communication. It can also support your emotional intelligence by helping you pause, reflect, and find better language in difficult moments. Use it as a tool to strengthen your leadership, not to replace your instincts. Parent leaders do not need to know everything. What matters is the willingness to keep learning, to use every resource available, and to grow alongside your children.

So, let's learn together. Welcome to *AI Powered Leadership for Modern Families*, where parenting meets purpose and technology finally works for you. Inside these pages, you will find more than two hundred AI prompts built around the ABCD Framework, tools designed to help you create healthier habits, stronger relationships, and a family culture grounded in empathy, accountability, and love. We are practicing this together, and that is what real leadership looks like.

PART I
Smarter Parenting

Your Family's Leadership Algorithm

Smarter parenting begins with awareness, the ability to pause, think, and lead with intention. In this section, you will explore the ABCD Framework, a practical guide that helps you operate as the Chief Family Officer, the person responsible for shaping your family's vision, values, and emotional climate. You will learn how to use artificial intelligence (AI) as a leadership tool that helps you clarify your thoughts, strengthen communication, and approach family life with calm, confident decision making.

AI does not replace intuition; it refines it. When guided by the ABCD Framework, AI becomes a tool for structured thinking and reflection. It can help you prepare for meaningful conversations, design family systems that bring order and balance, and guide your children, your Leaders-in-Training, toward emotional intelligence, curiosity, and accountability.

This is where leadership at home meets innovation. As the Chief Family Officer, you will learn how to use AI to turn everyday challenges into learning opportunities that grow both you and your children. Smarter parenting is not about having the answer; it is about having the question. When technology serves your values and your vision, your home becomes a place where structure, trust, and compassion thrive, and where every child learns what it means to lead with heart and wisdom.

1

CHAPTER 1

The Parent as CFO of the Family Enterprise

I n these pages, we will reimagine the family as a dynamic organization built on clarity of vision, purposeful leadership, and consistent communication. You will explore what it means to be the Chief Family Officer (CFO), the person responsible for setting the emotional tone, defining the mission, and aligning every member of the team around shared values. Each chapter transforms everyday parenting, from meltdowns to breakthroughs, into powerful lessons in leadership and growth.

You will learn how to:

- Define your family's vision and translate it into daily habits and growth.

- Establish guiding principles to shape decisions, discipline, and personal growth.

- Use communication as a tool that actually works and keeps your family focused, collaborative, and purposeful.

- Leverage the ABCD Framework to plan, reflect, and strengthen leadership behaviors within your family.

AI can also help you and your children become better students of both learning and life. It can support study routines, strengthen emotional intelligence, and spark curiosity through personalized questions, reflection prompts, and interactive activities. Using the ABCD Framework, you can guide AI to create meaningful conversations, organize family systems, and streamline daily routines. With this structure, you can operate your entire household more efficiently, saving time while maintaining calm and clarity. The same framework that supports thoughtful parenting can also manage meal planning, family meetings, and communication workflows with intention and ease. Children learn to stay motivated, reflect on progress, and take ownership of their responsibilities, while you model persistence, focus, and self-regulation. Together, you will learn to approach challenges with curiosity, use AI prompts built on the ABCD Framework to explore ideas, gather insights, and grow in understanding..

This is where the real transformation begins. You will move from managing your household to leading it with clarity and compassion. Great parents do more than raise children; they raise leaders who know how to think critically, feel deeply, and learn continuously.

How to Transform Your Parenting

Parenting can be accomplished by feeding your children, getting them to school on time, cooking dinner, and keeping laundry in motion. This approach meets basic needs, but parent leadership goes further. It involves raising children who are intellectually capable, emotionally mature, and confident in using technology responsibly. To achieve this, your children must first see a parent leader in action. This book will show you how to use AI as a tool to support the goals you have for your family, expands your creativity, and helps you lead with focus and calm.

Every day, whether you realize it or not, you already interact with technology. The goal here is to help you make it work for your family's benefit. You will learn how to pair leadership principles with the ABCD Framework so you can use AI to plan routines, solve problems, and practice the kind of reflection that strengthens emotional intelligence.

Parenting is an act of communication. Every word you speak, every tone you choose, and every question you ask teaches your child how to think and respond. The same principle applies when you communicate with AI tools. By using the ABCD Framework prompt, you will optimize the opportunity to clarify your intent, explore perspectives, and develop solutions aligned with your family's values.

Using the ABCD Framework will help you strengthen your child's emotional intelligence, sharpen their problem-solving skills, and teach them to handle challenges with clarity and confidence. Through simple, guided steps, you will learn how to shape AI prompts that help both you and your child become better students of life. AI can help organize study schedules, recommend creative ways to stay focused, and encourage healthy reflection after mistakes or conflicts. These experiences help children practice self-awareness, patience, and persistence.

You do not need to be a technology expert to begin. If you can send a text, use a search engine, or ask your phone a question, you already have the skills required. The ABCD Framework gives you a repeatable process for using AI as a family leadership tool. It helps you ask better questions, gather thoughtful answers, and create a culture of learning inside your home.

Understanding Large Language Models

Before you begin creating prompts, take a breath and imagine the playground you are about to enter. A large language model, or LLM, is simply an AI system that understands and responds to written language. You type a question or instruction, and it generates ideas, explanations, or reflections in return. The ABCD Framework will give you a clear structure to guide that process.

Technology may seem complex, but it operates on patterns just like parenting does. When you repeat bedtime routines, establish household expectations, or use familiar language to calm a tense moment, you are creating patterns your children learn to follow. Large language models do something similar. They learn from millions of examples of human communication to recognize patterns of tone, emotion, and reasoning. They do not feel, but they can help you think.

When used with intention, these models help parents lead more thoughtfully. They give you the time and mental space to pause, organize your thoughts, and approach each situation with empathy. When you apply the ABCD Framework, which includes the Actor, Background, Call to Action, and Data Check, you move from reacting to leading. You are teaching your children that even complex tools can be guided by clarity, compassion, and purpose.

Step-by-Step Instructions on how to use the ABCD Framework in a LLM

STEP 1: Choose a LLM

Begin by choosing a trusted large language model that feels comfortable for your lifestyle. Each platform offers a free version for exploration and optional upgrades to access more features. The objective is to select one you can access quickly and use often, just like a notebook or journal. A few reliable options include:

- **ChatGPT (OpenAI):** A well-rounded, conversational tool that can help with problem-solving, reflection, and emotional coaching.

- **Microsoft Copilot:** Built into Word, Edge, and other Microsoft tools; great for parents who live in their email or documents.

- **Claude (Anthropic):** Especially good at thoughtful, nuanced responses for emotional topics.

- **Perplexity:** A fast, research-oriented assistant that combines conversational answers with real-time web sources and is ideal for parents who enjoy data, facts, and staying current.

STEP 2: Open a New Chat

When you open a chat window in a LLM, you will see a blank space waiting for your words. That space is your Leadership Command Center. It is where the **ABCD Framework** comes to life. I created the **ABCD Framework** to give parents a reliable structure for using AI with confidence. It works like a blueprint to help you focus your thoughts and stay aligned to your family's needs. You will find more than 150 AI prompts for parents and children throughout this book, created to help you practice leading with curiosity, compassion, and confidence as you work with AI.

STEP 3: Lead with the ABCD Framework

Leadership begins with clarity, not perfection. The more direction you provide, the more meaningful the results will be. To begin, identify a prompt that relates to an experience you are currently having with your child and a large language model of your choice.

Next, copy and paste your AI prompt directly into your chosen large language model, such as ChatGPT, Copilot, or Claude. When you open one of these tools, you'll see a blank space waiting for your words. The prompt opens the conversation and gives technology a direction to follow. It is your way of saying, "Here's what I am working on; give me ideas to consider."

ABCD Framework

- **Actor**

- **Background**

- **Call to Action**

- **Data Check:** If you need more information, ask me questions before proceeding.

  A. Notice the ABCD Framework starts with the **Actor**, which refers to the lens through which you want AI to address the situation. The Actor can be a psychologist, coach, teacher, or another kind of expert. Taking the time to thoughtfully name the Actor helps the LLM understand whose point of view you want to explore.

B. Next, expand on the **Background** section of the prompt by adding more descriptive information about your specific situation. The more relevant details you include, the more precise and personal the response will be. For example, if the prompt asks about "setting morning routines," you might add that your eight-year-old struggles with transitions or that your teenager prefers music in the morning. Provide enough detail to help the model understand the situation. These small touches will replace a general response with a tailored leadership plan.

C. Now, we are tapping into your leadership on the **Call to Action**. This is the outcome that you want to create. It might sound like, "Develop a calm plan for homework time," or "Create a family conversation that encourages cooperation." The clearer your call to action, the more intentional and actional the response will be.

D. Finally, close with your **Data Check**. This is your moment to think like a leader and confirm whether your AI prompt includes enough information to generate a clear, effective response. The Data Check helps you confirm that your direction is thoughtful, complete, and aligned with the results you want to achieve. Include, "If you need more information, ask me questions before proceeding."

STEP 4: Select "Enter"

And then press, "Enter." Within seconds, your customized response will appear, often more thoughtful than you expect. Sometimes the AI will answer immediately, and other times it may ask a few clarifying questions to better understand your intent. This is part of the magic. It is learning with you, not just from you. Each exchange helps refine your leadership voice as you guide your family with greater focus on empathy. As you grow more comfortable, begin creating more and more prompts within the ABCD Framework.

If something feels incomplete, adjust your prompt. Say things like, "Make the tone more compassionate" or "Add an example that relates directly to my child's behavior." Keep iterating, meaning, refining and rewording your message, until it feels clear and aligned with your intent. You can always add more context in the Background or adjust your directive in the Call to Action to strengthen your results. Over time, this process becomes second nature, shaping the way you lead, reflect, and respond in every area of your life.

CHAPTER 2

The ABCD Framework

You do not need a technology degree or perfect spelling to use AI effectively. The ABCD Framework works seamlessly with all major AI tools, making it easier for parents to think, plan, and lead with intention. Think of these tools as digital assistants, available whenever you are ready to reflect, organize, or gain a fresh perspective. Whether you are typing between meetings or speaking quietly into your phone while folding laundry, AI can serve as a practical space to pause and think before you act.

Here's what it might look like in action:

Say this out loud to your AI app, "Actor: You are a child psychologist. Background: My ten-year-old just had a meltdown about homework. I want to stay calm and help them plan. Call to Action: Provide me with an age appropriate script to talk about this. Data Check: If you need more information, ask questions before proceeding."

It does not matter whether you type or speak your AI prompt; what matters is how you structure it. The ABCD Framework gives your request focus and context, much like entering the right destination before starting your GPS. When you use it correctly, your responses will be more relevant, practical, and emotionally aware.

At this point in the book, you already know the ABCD Framework: Actor, Background, Call to Action, and Data Check. Now, it's time to put it to work. Use the same structure within your chosen AI tool to guide your conversations and problem-solving. The clearer your input, the more valuable the response..

AI is not here to replace your instincts or your leadership; it exists to support your growth as a reflective parent. It cannot tuck anyone into bed or find the missing shoe, but it can help you think before you speak, plan before you act, and communicate in ways that build connection instead of conflict.

Think of AI as a tool to amplify your leadership voice. It helps you organize your thoughts, clarify your goals, and communicate with greater empathy and precision. When used with the ABCD Framework, AI becomes a structured way to strengthen both your clarity and composure. Each time you practice this process, you are not only leading your family with purpose and patience, you are also showing your children how emotionally intelligent leaders think carefully, speak intentionally, and respond with understanding.

When you use AI intentionally, the ABCD Framework becomes a habit of thoughtful leadership. You can use it to rehearse a difficult talk, prepare for a teacher conference, or simply organize your thoughts before a busy day. Each time you create or refine a prompt, you are modeling curiosity, flexibility, and self-awareness, the qualities of strong leaders and emotionally healthy families.

Start small. Try one prompt. Reword it until it feels natural. Let AI help you explore new perspectives and test ideas. Over time, you will begin to notice you are communicating with more patience and clarity, not only with technology but also with the people you love most.

Remember, great leadership is built on reflection and iteration. Every time you refine a prompt, you are also refining your mindset. You are showing your children growth comes from learning, not from perfection. And that is exactly what this book, and your journey as a parent leader, is all about.

Why the Four-Part ABCD Framework Works

This four-part structure doesn't just make AI smarter; it makes **you** smarter. Every time you use it, you are practicing critical thinking, clear communication, and self-reflection.

You are also modeling the exact leadership skills your child needs:

- **Emotional regulation**: staying composed under pressure.

- **Assertive communication**: using words wisely.

- **Leadership Mindset**: staying curious, open to feedback, and willing to grow.

You are showing your kids leaders don't just react; they prepare, reflect, and respond with intention. The best part? You are also learning.

An Example Prompt in Action

Here's what a full ABCD Framework prompt looks like in real life:

Prompt 1 - Parent: Growth Mindset on Homework

Actor: You are a child psychologist.

Background: My 12-year-old gets frustrated with his math homework and gives up easily. I don't want to just tell him to try harder.

Call to Action: Help me figure out how to talk to my son in a way that's supportive and encourages him to stick with it.

Data Check: If you need more information, ask me questions before proceeding.

By structuring your prompt this way, you create a genuine coaching session. AI becomes your sounding board (not your substitute) helping you gather your thoughts and lead with calm authority.

Here's where the gold really is. AI can't read your mind; it only knows what you share. When you invite it to ask clarifying questions, you open the door to deeper, more specific guidance. Doing a data check by telling AI, "If you need more information, ask me questions before proceeding," is the *wonder* tool. This is where you invite inquiry in case your prompt is not clear or complete enough to produce a response tailored to your specific situation.

It's like having a friend who doesn't just say, "Here's what to do," but instead asks, "Wait, what's really going on? Is it the mornings? The homework? The emotional overload?" Encourage your kids to do the same: ask questions before reacting. Encourage them to ask themselves, "Why am I frustrated?" and "Why do I feel rushed?" It's a leadership move. The more they practice curiosity, the less they'll jump to conclusions, and the more they'll learn to lead themselves.

Common sense still rules! AI is not raising your kids; you are. The tools presented are to support you. They are not a substitute for thoughtful parenting. You know your child's personality, triggers, and strengths better than any algorithm. Remember to use good judgment, check your gut, and seek professional advice when needed. Common sense still outranks artificial intelligence every time.

Tip: Try using these prompts before sitting down with your child. You'll feel calmer, clearer, and better equipped to lead the conversation instead of "winging" it. Preparation is half the battle and modeling calm leadership teaches your child what emotional maturity looks like in real life.

Ready to Begin?

This book includes over two hundred carefully designed prompts and covers everything from screen-time battles to self-esteem boosts to those big 'What am I doing wrong?' parenting moments.

Use them straight from the page or customize them for your specific needs. Every prompt helps you build connection, structure, and confidence while modeling problem-solving and reflection. Resist the temptation to resort to the Call to Action and leave out the remainder of the prompt. You and your family deserve the best.

Now, open the book. Pick an AI prompt embedded in the **ABCD Framework**. Start leading your family like the capable, loving, and occasionally sleep-deprived parent leader you are.

Because parenting with intention (and just the right amount of AI) requires leading with humor, humanity, and a touch of digital wisdom, one thoughtful prompt at a time.

CHAPTER 3

Smarter Together: How AI Strengthens Parent Leadership

Y ou are the leader of your home, the strategist, coach, and culture builder who sets the tone and rhythm for family life. While technology continues to accelerate the pace of the world, your children's deepest needs remain the same. They still depend on your wisdom, steadiness, and presence. Artificial intelligence can strengthen those qualities, but it can never replace them. When used with intention, you and AI become smarter together.

Parenting today involves managing complex and shifting demands. Modern parents face a steady flow of changing behavioral challenges, and the article suggests that AI conversational agents can deliver timely and adaptable guidance that supports families more consistently than traditional once a week sessions (Escoredo, Mostovoy, Schickler, Bechtel, Shagan, & Bunge, 2025). Further, AI can support the familial roles by providing real-time feedback, reminders, and other useful content (Szondy & Magyary, 2025). This support can stabilize family life even when challenges are constant.

Research between 2024 and 2025 confirms these benefits. Parents assume several roles such as teacher, collaborator, observer, and resource provider in support of their child's development of AI literacy (Druga, Bickmore, Hinniker, Vu, Likhith, & Qui, 2022). There are significant benefits of integrating AI into family systems such as promoting family cohesion (Szondy & Magyary, 2025).

AI can mirror caring language and an understanding tone, but it cannot build real emotional bonds or take the place of the love that comes from a parent. When your child seeks guidance about friendship, fairness, or values, they need your authentic responses. AI amplifies your instincts rather than acting as a co-parent. Adding your family's unique voice to AI-generated stories or solutions brings warmth and personal meaning, making learning experiences more genuine and lasting.

Studies also highlight AI's role in developing emotional intelligence within families. When families use AI reflection tools, they practice naming emotions, understanding triggers, and expressing themselves with clarity, all of which are essential skills for emotional intelligence (Escoredo et al., 2025). AI systems can detect shifts in tone and behavior, helping parents respond thoughtfully rather than react impulsively. AI can also analyze communication patents and identify miscommunication (Szondy & Magyary, 2025). Yet empathy and compassion remain human strengths that parents must nurture intentionally.

Preparing children for an AI pervasive world includes teaching AI literacy, which means engaging critically, ethically, and reflectively with AI. Parents model this literacy by explaining their reasoning when consulting AI and showing evaluation rather than blind acceptance. AI-generated ideas should be considered as suggestions and starting points for discussion.

Parental leadership also requires awareness of AI's biases and privacy concerns. AI systems learn from massive, often biased datasets and may reinforce stereotypes or overlook cultural nuances. Including family context and values in the AI Prompts embedded in the ABCD Framework will produce more authentic and relevant results. Parents are also responsible for teaching children about data privacy, digital responsibility, and the importance of cautious online behavior.

Clear communication sits at the heart of every strong and thriving family. The use of AI can augment healthy family dynamics through communication training and conflict resolution strategies (Szondy & Magyary, 2025). Still, AI cannot replace the relational work that defines your role. It can help you stay organized, but it cannot rebuild trust or create peace. The human heart remains the most powerful system in any home.

Leading with empathy, presence, and vision transforms every AI interaction into a teaching moment. Modeling curiosity, reflection, and thoughtful choices demonstrates how to balance power with compassion, which is the essence of leadership. Families that use technology wisely rather than excessively build trust and understanding through intentional, human-centered engagement.

Used thoughtfully, AI strengthens what already works within you, such as your instincts, patience, compassion, and humor. Combined with emotional intelligence, it becomes a mirror that reflects your best leadership back to you. Every heartfelt interaction proves that wisdom and warmth will always matter more than any algorithm.

You are the leader of that mission. You and AI are smarter together

AI Prompts for Parents

Prompt 2 - Parent: AI Protocol

Actor: You are a psychologist specializing in providing integrative care for family conflicts, anxiety, stress, and depression.

Background: I am the parent leader of my family. I want my family's digital presence to sound calm, respectful, and values-driven, even when life feels stressful.

Call to Action: Help me clarify the core values and beliefs that should guide every AI response in our family interactions so our use of technology reflects kindness, respect, and emotional growth., even when I'm stressed or multitasking.

Data Check: If you need more information, ask me questions before proceeding.

Prompt 3 - Parent: Establishing an AI Voice

Actor: You are a psychologist specializing in holistic, evidence-based family therapy practices.

Background: I am the parent leader of my family. I want every AI interaction in my family, planning meals, studying, or solving problems, to reflect the same perspective I use when guiding my children.

Call to Action: Help me identify the perspective through which I want all AI responses to my family's AI prompts.

Data Check: If you need more information, ask me questions before proceeding.

Prompt 4 - Parent: Managing AI Interactions

Actor: You are a psychologist specializing in solution-focused therapies for a broad spectrum of family concerns.

Background: I am the parent leader of my family. Our family's traditions, faith, humor, and shared experiences make us unique. I want that richness to come through in every AI response.

Call to Action: Help me define a clear family leadership perspective that guides all of our AI interactions. Every response, from meal planning to problem solving, should reflect our core values of empathy, respect, and growth. I want AI to model the same emotional intelligence, calm communication, and cultural awareness I strive to teach my children every day.

Data Check: If you need more information, ask me questions before proceeding.

Reflection AI Prompts for Young Leaders-In-Training

Prompt 5 - Student: Learning Approaches

Actor: You are a learning coach who helps kids learn in ways that make sense for their age.

Background: I am in the [enter here] grade and I am [enter here] years old. I like learning through stories, games, and real-life examples that make learning fun.

Call to Action: Please suggest fun and encouraging ways to approach learning new or challenging school work.

Data Check: If you need more information, ask me questions before proceeding.

Prompt 6 - Student: Critical Thinking

Actor: You are a smart online guide who helps kids like me figure out what's real and what's not on the internet.

Background: I am in the [enter here] grade and I am [enter here] years old. I know some things online are neither fair or true, and I want to learn how to recognize bias or dangerous content.

Call to Action: Can you teach me how to evaluate information and determine if it could be biased or trying to take advantage of kids?

Data Check: If you need more information, ask me questions before proceeding.

Prompt 7 - Student: Emotional Intelligence

Actor: You are a coach who helps kids like me learn how to talk so adults really get what we mean.

Background: I am in the [enter here] grade and I am [enter here] years old. I want to learn how to communicate so my parents will understand and respect my voice.

Call to Action: Can you help me customize AI so when I am using AI prompts the responses will help me learn how to be respectful, kind, and clear?

Data Check: If you need more information, ask me questions before proceeding.

PART II
Leading Through the Real-World Challenges

You've built the foundation, now comes the leadership lab.

The first chapters helped you think like a parent leader: setting the emotional tone, shaping your family culture, and using AI as your personal coaching tool. But leadership doesn't live in theory; it shows up in the tension between homework and Wi-Fi, between a slammed bedroom door and the quiet hope tomorrow's conversation will go better.

This next section is where your leadership gets tested and strengthened. You'll step into real-life parenting challenges, the ones that make you question your patience, your choices, and sometimes your sanity. Each chapter offers a scenario, a leadership reflection, and an opportunity to practice your ABCD prompting framework. Think of it as your leadership practice: real problems, real emotions, and real growth.

As you read, pause to ask:
What is this moment teaching me about leadership at home?
How can I model composure, accountability, and empathy right here?
What prompt could help me reflect before I react?
This is where the theory becomes practice.
This is where parenting becomes leadership in motion.
Welcome to the heart of the work - the messy, meaningful middle.

CHAPTER 4

Resistance to Increasing Academic Effort

"My daughter's test grades are beginning to decline. But she refuses to spend more time studying and won't take responsibility for her performance. She makes up stories about how her teacher doesn't want her to work ahead, study more, or do extra problems. When I confront her, she cries and becomes defiant. What can I do?"

It often begins with something small. A quiz score that feels lower than usual. A teacher's note about needing more practice. A math grade that makes you look twice. You bring it up gently at dinner, without judgment, simply suggesting that a little extra review might help. Then suddenly your child gives a detailed explanation about how the teacher discourages extra work. The tears come, emotions rise, and dinner becomes tense.

What you are witnessing is not laziness. It is fear in a costume. When children encounter real academic challenges for the first time, effort can feel unfamiliar and uncomfortable. If success has come easily, working harder can feel like proof they were never capable in the first place. When fear meets pride, resistance shows up.

This is the moment for you to lead, not to lecture. You are guiding a young learner through the emotional landscape of growth. The first leadership step is setting the emotional tone. Your calm energy is the thermostat for the entire household. When your child's frustration fills the room, take a slow breath and speak gently. Say, "I can see this is hard for you. You are capable, and we can figure this out together." This simple statement communicates safety and possibility. It reminds your child a bit of struggle is not the end of the story. It is where growth begins.

Once calm returns, clarify expectations and boundaries. These are not punishments. They are promises to help your child feel supported and secure. Boundaries provide structure, and structure builds confidence. Try saying, "In this family, we prepare and give our best effort before we decide something cannot be done." Consistent routines teach responsibility, while compassionate follow-through teaches trust.

The next leadership step is teaching and modeling problem solving. Children learn more from how we handle challenges than from what we tell them. Share a story about a time you struggled and recovered. Describe what you learned and how you grew. Let your child see that adults also practice perseverance.

This is a wonderful opportunity to use technology as a learning tool. Invite your child to use AI as a tool rather than a shortcut. Sit together and model how to ask for help in a thoughtful way using the ABCD Framework:

Prompt 8 - Parent: Process over Perfection

Actor: You are a kind tutor who helps middle school students understand math.

Background: My 9 year old daughter is learning fractions and feels unsure about how to begin.

Call to Action: Create a plan with examples and ways to practice.

Data Check: If you need more information, ask me questions before proceeding.

This process builds confidence and curiosity. It transforms technology into a tool for learning rather than a source of avoidance. You are showing your child that asking for guidance is strength, not weakness.

As you move forward, hold your child accountable with empathy. Accountability without compassion creates resistance, while compassion without accountability creates confusion. Balance both. When your child blames the teacher or avoids responsibility, listen first. Then ask, "What part of this do you think you can control?" This question shifts the focus from excuses to empowerment. When you meet with the teacher, model calm collaboration. Say, "We want to understand how to best support her learning at home." You are teaching your child the importance of handling difficulty through communication, not confrontation.

Continue nurturing a growth mindset and a vision for the future. Remind your child progress matters more than perfection. Say, "We learn by practicing. Every effort moves you forward." Celebrate persistence, curiosity, and creativity. Ask reflective questions such as, "What did you learn about extra effort while you were studying?" or "What felt easier this time?" These questions build self-awareness and reinforce the value of steady effort.

Sometimes resistance signals something deeper. A child who avoids work may be struggling to: see the white board in class, manage anxiety, or organize thoughts. Curiosity is your greatest leadership tool. Ask, "Is something else making this harder?" You might uncover an obstacle at the crux of the attitude. Get the vision or hearing check. Seek guidance from a counselor or tutor if needed. Reaching out for help is not giving up; it is a wise and loving step toward success.

Remember, your child's homework is not your homework. You are the coach, not the player. Your role is to provide encouragement, structure, and accountability while allowing your child to build independence. When he says, "I do not know how," respond with, "Let's figure it out together." This approach communicates faith in his potential and teaches problem solving through partnership.

The true goal is not perfect grades. It is the ability to think, adapt, and persevere. You are raising a young person who can say, "I do not know yet, but I can learn." This mindset will carry your child through every classroom, every workplace, and every life challenge to come. You are not running a homework factory. You are leading a family that values curiosity, growth, and courage.

Parenting leadership means guiding with calm, clarity, and compassion even when resistance appears. Each moment of frustration is an invitation to practice patience, communication, and emotional intelligence. You are teaching your child not just how to study but how to believe in their capacity to learn and grow.

Reflection Questions for Parents

- How do I respond when my child resists studying or learning?

- What emotional tone do I set during moments of frustration?

- Are my expectations clear and consistent while remaining kind?

- How can I demonstrate problem solving and patience in my own actions?

- What messages about growth and effort do I want my child to carry into adulthood?

Reflection AI Prompts for Parents

Prompt 9 - Parent: Clarity is Kindness

Actor: You are a Parenting Coach who specializes in emotional regulation and growth mindset.

Background: My child is in the [enter grade] grade. I get frustrated when my child resists studying and I want to react more calmly.

Call to Action: Please help me reflect on my own emotions and develop strategies to stay calm while keeping clear expectations. Help me uncover my own fears underlying my frustration.

Data Check: If you need more information, ask me questions before proceeding.

AI Prompts for Parents

Use these prompts to get actionable strategies for supporting your child effectively:

Prompt 10 - Parent: Homework and Process

Actor: You are an experienced teacher who helps parents support learning at home.

Background: My child is in the [enter grade] grade and they get overwhelmed by big assignments.

Call to Action: Please give me strategies to help them break work into manageable steps.

Data Check: If you need more information, ask me questions before proceeding

Prompt 11 - Parent: Homework and Emotional Intelligence

Actor: You are a family counselor who helps create peaceful homework routines.

Background: My child is in the [enter grade] grade. Homework time is stressful and full of arguments.

Call to Action: Please suggest ways to make it calmer and more positive.

Data Check: If you need more information, ask me questions before proceeding.

Prompt 12 - Parent: Growth Mindset Leadership

Actor: You are an educational psychologist who explains mindset in kid-friendly terms.

Background: I want my [insert grade] grader to understand the growth mindset.

Call to Action: Please help me explain it in simple words.

Data Check: If you need more information, ask me questions before proceeding.

AI Prompts for Young Leaders-In-Training

Teach your child to use these themselves so they learn to lead their own learning with your guidance:

Prompt 13 - Student: Homework and Process

Actor: You are a friendly tutor who helps kids study.

Background: I am [enter here] years old and I am in the [enter here] grade. I have a test coming up and don't know how to prepare.

Call to Action: Please help me make a plan to study for it.

Data Check: If you need more information, ask me questions before proceeding.

Prompt 14 - Student: Homework Guidance

Actor: You are a learning coach who helps with comprehension.

Background: I am [enter here] years old and I am in the [enter here] grade. I don't understand something I'm reading or learning.

Call to Action: Please give me questions to ask myself to figure it out. I am working on [insert here] and I am stuck on [enter here].

Data Check: If you need more information, ask me questions before proceeding.

Prompt 15 - Student: Homework and Process

Actor: You are a study skills coach for students.

Background: I am [enter here] years old and I am in the [enter here] grade. I want to take better notes that actually help me remember.

Call to Action: Please teach me how to take effective notes on this subject: [enter here].

Data Check: If you need more information, ask me questions before proceeding.

CHAPTER 5

Building Strong Study Habits - Leading Lifelong Learning

"After a long day at work, I walked into my house and my 12-year-old was sprawled on the sofa watching TV with his bookbag forgotten at the door. How many times do I have to tell him to do his homework first?"

If you have ever come home after a long day, hoping for a quiet moment, only to discover your child has not even opened a book, you are in familiar company. Maybe they are stretched out on the couch, glued to a screen, or wandering around the house insisting there is "no homework." Perhaps you have fought this battle so many times you feel weary before the conversation even begins. You ask, they resist, and the evening ends with both of you frustrated and discouraged.

As parents, we want our children to succeed. We want them to value education, take pride in their work, and develop habits that will serve them for life. Yet it can feel as though we are the only ones who care. We wonder why they will not take responsibility or manage their time, and we grow tired of repeating ourselves. Beneath our frustration lies love and hope. We want them to care as deeply as we do.

Here is an important truth to hold with empathy: children are still learning how to learn. The part of their brains that manages planning, time, and focus is still developing well into adolescence. What appears to be laziness or disinterest is often an unfinished skill needing structure, guidance, and repetition. This is where leadership at home begins. The way we respond to our children's learning moments shapes how they see themselves as learners. Emotional intelligence must come before action. When we manage our own frustration, we model what self-regulation looks like. When we stay calm under pressure, we create an atmosphere where growth feels possible.

Parenting is leadership. It is not about commanding obedience rather it is about guiding development. We set the emotional tone of the home. Our children watch how we handle pressure, how we speak when we are tired, and how we recover when we lose our patience. Every sigh, every word, every gesture teaches them something. When we approach homework time with empathy instead of irritation, we teach discipline is not punishment; it is a rhythm of care. Calm leadership says, "You are capable, and I will help you learn how to succeed."

You do not have to act as if homework time is delightful, but you can choose to lead it with composure. Try saying, "Let's both take a few minutes to unwind, then we can focus together." Sit beside your child while they work. Read, answer emails, or plan your day nearby so they see learning as something the family values, not just something children must do. The energy you bring into these moments communicate more than any rule you enforce. When they see you steady and engaged, they learn difficult things can be managed with patience.

Sometimes the best leadership happens before the conversation even begins. Take a few moments after work to gather yourself before stepping into homework time. A pause, a deep breath, or even a few minutes of silence can transform the tone of the entire evening. When you enter calm, you teach calm. When you apologize after losing your temper, you teach humility. Children learn far more from how we recover than from how we react.

Boundaries and structure are not barriers to freedom; they are pathways to trust. When children know what to expect, they feel safe. A predictable routine prevents nightly negotiations about when homework begins or how long screens can stay on. A simple rhythm such as a snack, study, then free time helps children understand structure is a form of love. It communicates that your family has shared values and everyone's time matters. Even the way you prepare the study space sends a message. A clean, well-lit table, a comfortable chair, and a healthy snack says, "Your work matters, and I believe in your ability to handle it." Routines build reliability, and reliability builds confidence.

As parent leaders, we teach by modeling problem solving. When your child becomes frustrated, resist the urge to fix it right away. Instead, help them think through what to do next. This is where technology can serve as a helpful tool. AI can assist with organizing assignments, creating practice problems, or breaking large projects into smaller steps. When you show your child how to ask a thoughtful question or check their understanding, you are modeling how to use technology as a support system, not a crutch. A prompt like, "You are a patient tutor who helps students understand fractions. Create a simple plan for how I can practice," shows your child learning is about curiosity, not shortcuts. The goal is not to let AI think for them but to help them think better.

Accountability becomes meaningful when it is paired with empathy. If your child forgets an assignment or blames the teacher, begin by listening. Then ask, "What part of this do you have control over?" That question shifts the conversation from blame to ownership. Consequences work best when they stay connected to the choice that caused them. Saying, "That grade reflects the effort you put in. Let's plan what you'll do differently next time," turns a setback into a learning opportunity. Accountability is not about embarrassment or shame; it is about teaching responsibility with respect.

Empathy does not mean giving in. It means holding firm boundaries with compassion. You might say, "I know you are tired. I am too. But this is our study time, and I know you can get through it." This combination of warmth and structure builds trust. It reassures your child discipline is not rejection; it is preparation for adulthood. Even when mistakes happen, they can become lessons in resilience. When children understand errors are not evidence of failure but opportunities for insight, they begin to take pride in their effort rather than only their results.

Parent leadership also means nurturing a growth mindset and connecting today's habits to tomorrow's freedom. Help your child see that learning is not a chore but a lifelong advantage. Say, "The time you spend learning how to manage your schedule now will help you later in high school, in college, and even when you have your own family." Inspire them to think beyond the homework in front of them. Every small act of consistency is an investment in their future independence.

Celebrate progress, not perfection. When your child starts homework without being asked, notice it. When they show persistence after struggling, acknowledge their effort. Words like "You stayed with it" or "You planned that well" reinforce internal motivation. Let them see that success is built on steady practice, not natural talent. Share your own experiences of learning something new or solving a problem with technology. Say, "I do not know the answer, but let's find out together." This communicates growth is not limited to children; adults are still learning too. You are showing curiosity never expires.

When we lead our families with emotional intelligence, clear expectations, empathy, problem solving, and a belief in growth, we do more than build good students, we shape good thinkers. The goal is not to control behavior but to develop character. Every calm tone, every consistent boundary, every respectful correction plants seeds of resilience, responsibility, and confidence. The structure you create today will become the self-discipline your child carries into adulthood. The patience you practice will become their inner calm when life tests them. And the faith you show in their potential will echo in the way they lead their own lives tomorrow.

Reflection Questions for Parents

- How do I set the emotional tone during homework time?

- Are my expectations clear and consistent?

- Do I teach my child how to study or simply tell them to study?

- How do I balance accountability with empathy?

- How can I help my child see learning as valuable for their future?

AI Prompts

- ✓ Suggest one change to strengthen my child's homework routine this week.

- ✓ Provide a script for me to use to talk to my child about why learning matters.

- ✓ How can I create a calm, welcoming study space?

- ✓ Provide suggestions on how I can model my own learning and self-discipline through everyday actions.

Prompt 16 - Parent: Make Homework Time More Positive

Actor: You are a Family Coach who specializes in positive parenting.

Background: My child is in the [insert grade here] grade. Homework time in our house often turns into nagging or arguing. I want it to feel calmer and more productive.

Call to Action: Give me ideas to make our family's homework time more positive.

Data Check: If you need more information, ask me questions before proceeding.

Prompt 17 - Parent: Write a Calm, Encouraging Script

Actor: You are a Parenting Coach who helps write gentle, effective scripts.

Background: My child is in the [insert grade here] grade. I want to talk to my child about study expectations without sounding mean or bossy.

Call to Action: Provide me with a calm, encouraging script for talking about study expectations.

Data Check: If you need more information, ask me questions before proceeding.

Prompt 18 - Parent: Balance Rules with Empathy

Actor: You are a Family Counselor experienced in setting boundaries with empathy.

Background: My child is in the [insert grade here] grade. I want to enforce our homework rules but also show my child I understand how they feel.

Call to Action: How can I balance enforcing rules with showing empathy?

Data Check: If you need more information, ask me questions before proceeding.

Prompt 19 - Parent: Create a Weekly Study Schedule

Actor: You are an Educational Coach for busy families.

Background: Our schedule is packed with work and activities. I want a realistic weekly study routine. My child is in the [insert grade here] grade.

Call to Action: Suggest a weekly study schedule for busy families.

Data Check: If you need more information, ask me questions before proceeding.

AI Prompts for Young Leaders-In-Training

Show your child how to **use AI prompts for learning.** These prompts help them plan, reflect, and get unstuck, building independence and problem-solving skills.

Prompt 20 - Student: Break a Project into Steps

Actor: You are a kind teacher who helps kids plan projects.

Background: I am [enter here] years old and I am in the [enter here] grade. I have a big project due and I feel overwhelmed. This is the assignment: [Insert assignment here]

Call to Action: Help me break this project into steps.

Data Check: If you need more information, ask me questions before proceeding.

Prompt 21 - Student: Make a Homework Checklist

Actor: You are a homework helper for kids.

Background: I am [enter here] years old and I am in the [enter here] grade. I forget steps and lose track of what to do first. This is the assignment: [Insert assignment here]

Call to Action: Give me a checklist to finish my homework.

Data Check: If you need more information, ask me questions before proceeding.

Prompt 22 - Student: Get Unstuck on a Question

Actor: You are a friendly tutor who helps kids think through challenging problems.

Background: I'm stuck on a question and don't know what to do next. This is the question: [Insert question here]. I am [enter here] years old and I am in the [enter here] grade.

Call to Action: Teach me how to solve these problems without looking up the answers.

Data Check: If you need more information, ask me questions before proceeding.

Prompt 23 - Student: Remember Assignments

Actor: You are an organizational coach for students.

Background: I am [enter here] years old and I am in the [enter here] grade. I keep forgetting to write my homework down.

Call to Action: How can I remember to write my assignments down?

Data Check: If you need more information, ask me questions before proceeding.

Prompt 24 - Student: Learn About A Growth Mindset

Actor: You are a teacher who explains big ideas in kid-friendly ways.

Background: I am [enter here] years old and I am in the [enter here] grade. Sometimes I get really upset when I make mistakes on homework.

Call to Action: Explain growth mindset for kids.

Data Check: If you need more information, ask me questions before proceeding.

CHAPTER 6

Helping Your Young Leader Stay on Top of Homework

"Last night we were up until 1 a.m. finishing a project my daughter 'forgot about.' I was exhausted from work, and she was in tears. How did we get here?"

I f this sounds familiar, you are not alone. Many parents know the sinking feeling of hearing ,"I have a project due tomorrow" just as they are about to collapse into bed. The day has already stretched you thin, you woke up early, got everyone ready, handled work, traffic, meals, and the million quiet tasks related to running a household, and now, at the end of it all, you are holding scissors, glue sticks, and frustration at the kitchen table!

When actively parenting, the days rarely end uneventfully. Evenings bring new demands: forgotten homework, sibling arguments, dinner dishes, late work messages, and the constant pull to show up for everyone all at once. The mental load is invisible but enormous. When your child wanders in at 9 p.m. with a last-minute project, your body may tense before your mind even catches up. You feel angry, anxious, and helpless. You want to help, but you also want to scream. You worry this pattern will repeat forever.

The worry comes from fear. We want our children to succeed, to care about their responsibilities, to learn how to manage their time. Beneath the frustration, we are afraid if they don't master this now, they will stumble later, through high school, college, and adulthood. But parenting, like leadership, begins with emotional intelligence before action. When your child's poor planning collides with your exhaustion, the best response often begins with a breath. A calm tone teaches more than a lecture ever could. You are not just helping with homework; you are modeling how to handle pressure, disappointment, and responsibility.

Your child will remember not only what you said but how you said it. They are studying your reactions as much as their own assignments. When you meet their panic with your own, the whole house absorbs that energy. But when you pause, steady yourself, and say, "We'll figure out a better plan next time," you are leading with emotional balance. The tone of composure is what leadership looks like in a family, it transforms chaos into a classroom for growth.

Leadership at home also requires clear expectations. Children thrive in predictable environments. They need to know when homework happens, where it happens, and what support is available. This doesn't mean standing over them or rescuing them every time. It means establishing family systems to prevent a crisis from taking root. You might say, "Homework time starts at seven," or, "Let's check the planner together before dinner." Consistency replaces chaos, and structure says, "Our family values preparation." When children know the rules, they can rise to meet them.

Still, even with structure, late-night surprises will happen. When they do, leadership asks you to balance empathy with accountability. You can hold boundaries without crushing their spirit. It sounds like, "I know this feels stressful, but it's part of learning to plan ahead. I'm here to help you think it through, not to do it for you." This type of response delivers both compassion and firmness. It communicates, "I believe in your ability to improve." Consequences still matter. Sleepy mornings and low grades are natural outcomes; but when delivered with empathy, you build trust instead of fear.

It is tempting, in those moments, to blame yourself. You may replay the evening in your head, "what could I have said differently"; when you should have asked about homework, how you lost your temper. Guilt is a familiar companion for parents. It whispers that you are not patient enough, not attentive enough, not organized enough. But guilt is not leadership. Reflection is. The difference lies in what you do next. Leaders learn from mistakes, including their own. When you apologize for shouting or admit you're tired, you model humility. When you calmly explain you're changing how you handle homework from now on, you teach adaptability. Children learn emotional intelligence by watching it in motion.

At the same time, part of leadership is understanding what's really going on beneath the behavior. Forgetfulness, procrastination, and last-minute panic often have less to do with defiance and more to do with underdeveloped executive functioning in the part of the brain that manages time, organization, and planning. Expecting flawless organization from a middle schooler is like expecting them to parallel park. They simply haven't mastered it yet. Recognizing this doesn't excuse chaos, but it reframes it. Instead of saying, "You're irresponsible," you can say, "You're still learning how to plan

47

ahead, and I'm going to help you build that skill." This single shift turns punishment into coaching.

Coaching means asking questions to build awareness: "What made it so hard to remember this assignment?" or "What could you do differently next time?" You are training your child to think critically about their actions. In the process, you are modeling problem solving. You are showing mistakes are not permanent, but steps to learning. This approach aligns with a growth mindset, the belief that ability can be developed through effort, practice, and feedback. Psychologist Carol Dweck's research confirms when children are praised for effort rather than innate talent, they develop resilience. They begin to see challenges as opportunities rather than signs of weakness.

Your language as a parent leader matters deeply. A fixed mindset says, "You're lazy" or "You never plan ahead." A growth mindset says, "You haven't mastered this yet, but you can." The word, *"yet"*, can change everything. It turns frustration into faith. It reminds your child growth is possible and expected. When you speak with curiosity instead of criticism, your child learns they are safe to make mistakes, reflect, and try again. They learn to see effort as strength, not struggle.

These principles also apply to the role of technology in your home. Rather than banning every distraction or letting screens dominate, teach your child to use technology as a tool for organization and learning. You can model this by saying, "Let's ask an AI tool to help us make a schedule for the week."

Use prompts like:

Prompt 25 - Parent: Project Planning

> **Actor**: You are a study coach who helps middle school students plan projects.
>
> **Background**: My project is due in five days, and I don't know how to start. I am [enter here] years old and I am in the [enter here] grade.
>
> **Call to Action**: Create a step-by-step plan with daily goals.
>
> **Data Check**: If you need more information, ask me questions before proceeding.

By demonstrating how to use digital tools intentionally, you are not just solving tonight's problem, you are teaching digital discipline and responsible self-leadership.

Leadership in parenting often means resisting the urge to rescue. It means guiding your child through discomfort instead of removing it. It means staying close but not taking over. You might say, "I'll stay nearby while you work on this, but it's your project." This communicates belief in their capacity. It teaches ownership. It helps them feel capable even when they feel overwhelmed. Your presence becomes a quiet signal of support, not control.

There will still be nights when tempers flare and patience runs thin. Leadership is not about avoiding conflict; it is about managing it with grace. When you lose your cool, repair it. When they melt down, stay steady. When the night ends with tears, wake up the next morning and say, "We'll do better together." A simple phrase to rebuild trust and reinforce mistakes, yours and theirs, are part of learning and growing.

Over time, this steady, emotionally intelligent leadership reshapes the entire rhythm of the household. You begin to notice fewer last-minute panics and more proactive planning. Your child starts to anticipate instead of react. They feel more confident and responsible because you have taught them, through tone, consistency, and example, how to lead themselves. They begin to understand accountability is not about punishment; it is about growth.

The ultimate goal is not perfect homework habits. It is raising a child who sees challenges as chances to improve, who believes effort leads to mastery, and who knows love and accountability can coexist. This is what leadership in a modern family looks like. It is not defined by control but by composure, not by punishment but by presence. When you show your child that learning happens in the messiness of real life, you are helping them build resilience far beyond the classroom.

So, the next time your child announces a forgotten project at bedtime, take a breath, remember your role, and lead with vision. You are not just helping them finish an assignment, you are shaping how they will face every challenge life brings. Calm is your strategy. Clarity is your structure. Empathy is your power. Growth is your shared goal. And that is the kind of leadership that lasts long after the homework is done.

AI Prompts for Parents

How to Use Your Prompt (Step by Step Example):

Prompt 26 - Parent: Empathetic Accountability

> **Actor:** You are a parenting coach who specializes in calm communication.
>
> **Background:** I get really angry when my child forgets their homework and I end up yelling. I want to handle this better.
>
> **Call to Action:** What can I say instead of yelling when I'm frustrated?
>
> **Data Check:** If you need more information, ask me questions before proceeding.

Other Parent Prompt Ideas for Different Call to Actions:

Reflect on Your Own Reactions

Use AI to help you understand your emotions and choose better responses.

Example full prompt:

Prompt 27 - Parent: Emotional Trigger Awareness

> **Actor:** You are a family counselor who helps parents address academic behavior calmly.
>
> **Background:** I feel furious when my child forgets homework over and over, and I know my reaction isn't helping.

Call to Action: Why do I get so angry when my child forgets homework?

Data Check: If you need more information, ask me questions before proceeding.

Other ideas:

- What can I say instead of yelling when I'm frustrated?

Plan Calm Conversations

Get help scripting difficult talks with your child in a caring, effective way.

Example full prompt:

Prompt 28 - Parent: Empathetic Accountability

Actor: You are a parenting coach who writes calm conversation scripts.

Background: My child struggles with planning and often forgets assignments. I want to talk about it without sounding mean.

Call to Action: Write me a calm script to talk with my child about planning better.

Data Check: If you need more information, ask me questions before proceeding.

Other ideas:

How can I explain the consequences of poor planning without sounding mean?

Help me identify relevant consequences to the behavior and strategies to enforce planning.

Design Better Systems

Use AI as a brainstorming tool for creating routines and tools that work for your family.

Example full prompt:

Prompt 29 - Parent: Homework Routine Planning

> **Actor:** You are an organizational coach for busy families.
>
> **Background:** We are always rushing and sometimes forget homework until bedtime.
>
> **Call to Action:** Give me ideas for a simple homework routine for a busy family.
>
> **Data Check:** If you need more information, ask me questions before proceeding.

Other ideas:

- Help me create a daily homework log template.

Strengthen Your Parenting Mindset

Use AI to reflect on the kind of leader you want to be for your child.

Example full prompt:

Prompt 30 - Parent: Growth Mindset Modeling

Actor: You are a positive parenting expert.

Background: I want to teach my child to see mistakes as learning opportunities.

Call to Action: How do I model a growth mindset for my child?

Data Check: If you need more information, ask me questions before proceeding.

Other ideas:

- List ways to show empathy while setting boundaries.

AI Prompts for Young Leaders-In-Training

To Plan and Break Down Assignments:

Show your child how to use AI to create structure and reduce overwhelm.

Example prompt:

Prompt 31 - Student: Project Planning Support

Actor: You are a kind teacher for 2nd graders.

Background: I have a poster project due next week, and I don't know how to start. I am [enter here] years old and I am in the [enter here] grade.

Call to Action: Please help me plan this project in daily steps.

Data Check: If you need more information, ask me questions before proceeding.

Other prompt ideas:

- What are good ways to measure my success as I work on my brochure?

- Make a checklist for finishing my project on time.

To Improve Work Quality

Help your child see AI as a helpful reviewer or advisor for making their work shine.

Example prompt:

Prompt 32 - Student: Creative Problem Solving

Actor: You are an art teacher who helps kids improve projects.

Background: I'm making a poster for school, but I want it to look better. I am [enter here] years old and I am in the [enter here] grade.

Call to Action: Suggest how I should approach creating an interesting poster.

Data Check: If you need more information, ask me questions before proceeding.

Other prompt ideas:

- What should I double-check before turning in a writing assignment?

- How can I make sure my project is neat and complete?

To Practice Reflection and Learning:

Model self-awareness and problem-solving after setbacks.

Example prompt:

Prompt 33 - Student: Learning Accountability

Actor: You are a school counselor who helps kids learn from mistakes.

Background: I forgot to turn in my homework yesterday. I am [enter here] years old and I am in the [enter here] grade.

Call to Action: Suggest strategies I could use to help me remember next time?

Data Check: If you need more information, ask me questions before proceeding

Other prompt ideas:

- How can I remember to write down homework every day?

- How can I get better at starting big projects early?

To Develop a Growth Mindset

Encourage your child to see challenges as opportunities.

Example prompt:

Prompt 34 - Student: Process over Perfection

Actor: You are a teacher who explains growth mindset to kids.

Background: I get really upset when I make mistakes on homework. I am [enter here] years old and I am in the [enter here] grade.

Call to Action: Please give me examples of how mistakes help me learn.

Data Check: If you need more information, ask me questions before proceeding.

Other prompt ideas:

- "Explain growth mindset for kids."

- "How can I stay calm when I feel overwhelmed by homework?"

Tip: Make this collaborative. Let your child help choose which prompt to try. This turns stressful moments into learning opportunities and reinforces your role as the leader who teaches, guides, and supports, not rescues.

When you use prompts embedded in all four parts of the ABCD Framework, you are doing more than solving the problem at hand; you are *teaching your child how to ask good questions, reflect thoughtfully, and use technology as a powerful learning resource.*

Action Steps

- Choose one or two of the parent prompts to try this week.

- Introduce one or two child-friendly prompts in your next homework talk.

- Make using AI a family routine for big projects, *before* the night-before panic.

CHAPTER 7

Reasoning with Rubrics

"My son and I frequently argue about the quality of his schoolwork. He claims he is doing quality work. But, his grades do not indicate quality. Honestly, his work is basic. How can I help him understand the importance of producing quality schoolwork?"

I f this sounds familiar, you are not alone. As the leader of your family, you manage school deadlines, chase missing assignments, and teach your child what it means to take pride in their work. It's not an easy lesson, especially when their idea of "good enough" is miles away from yours.

The truth is, most children don't actually know what "quality work" looks like. From their perspective, finishing at all feels like success. They haven't yet lived in the adult world where sloppy work means redoing the same thing twice, missing opportunities, or disappointing a client or supervisor. They do not see cutting corners is not efficient; it just moves the problem down the line. Your job is not to shame them for not knowing better. It is to make the invisible visible. Show them what good work looks like, feels like, and why it matters.

That's where rubrics come in.

When we hear the word *rubric*, many of us picture a red pen, a teacher's desk, and a graded paper covered in checkmarks and strikethroughs. But a rubric is simply a roadmap for quality. It defines what success looks like before the work even begins. It turns guesswork into clarity. In fact, we already use rubrics every day.

Think about it. When you follow a recipe, you are using a rubric: a clear list of ingredients, steps, and timing that leads to a predictable result. When you check hotel reviews before booking a trip, you are using a rubric comparing location, cleanliness, and service against what "excellent" looks like. Even when you buy groceries, you are quietly grading: the ripeness of a peach, the color of spinach, the freshness of milk. You know instinctively what earns a "4" and what drops to a "2."

In your workplace, rubrics show up in performance reviews, project goals, and customer satisfaction surveys. You might not call them rubrics, but every deadline, metric, or standard you follow is a rubric. When you expect a meeting to start on time, when you measure success by outcomes, or when you tell a coworker a report needs "more detail," you are referring to an internal rubric. These everyday rubrics guide our decisions, behaviors, and priorities. They are invisible grading systems that help us navigate life with consistency and accountability.

Children, however, often live without these invisible guides. They know what *done* means, but not what *excellent* means. For them, finishing an assignment feels like an "A", because their rubric only has one criterion: completion. The gap between their version of quality and the teacher's version is where tension begins.

This is why defining a rubric together is so powerful. It replaces vague criticism with shared understanding. It moves you from "just do better" to "let's define better together." In your family, think of it as your quality control playbook. Every organization, from schools to businesses, thrives on clear standards. The home is no different. When expectations are clear, there is less arguing, less confusion, and more pride in the outcome.

Start by gathering what already exists. Reach out to your child's teacher and ask for the rubrics they use for writing, math, or science. Treat them not as report cards but as conversation tools. Sit down with your child and walk through the criteria together. Ask, "What does a top score mean to you? How would you know you met it?" Talk through examples of what "excellent," "good," and "needs work" really look like. If they do not have a teacher-provided rubric, create one together. Keep it simple but specific.

For instance, here's what a sixth-grade writing rubric might look like:

- **Score 4 (Excellent):** Writing is clear, organized, and engaging. Sentences are complete, grammar is correct, and punctuation and spelling are accurate. Word choice is varied and thoughtful.

- **Score 3 (Proficient):** Writing is mostly organized and clear with only minor errors. Ideas are developed but may lack depth or variety in word choice.

- **Score 2 (Developing):** Writing shows effort but lacks structure or clarity. Errors distract the reader, and ideas are not fully developed.

- **Score 1 (Needs Improvement):** Writing is confusing or incomplete. Frequent errors make it hard to follow.

- **Score 0 (Incomplete):** Assignment is missing or not submitted.

This type of rubric doesn't just evaluate, it educates. It teaches the difference between effort and excellence. It helps your child clearly see quality is measurable, achievable, and within their control.

Here's the powerful part: rubrics don't just belong in the classroom. You can use them to teach life skills in almost every area of family life.

- **Morning Routines:** A "4" might mean getting ready on time, with everything packed and no reminders needed. A "2" might mean needing help finding shoes or forgetting a backpack.

- **Chores:** A "4" means the task was done fully and neatly. A "3" means it was done but rushed. A "1" means you had to remind them several times to complete the needed tasks.

- **Attitude:** A "4" could mean cooperating without arguing; a "2" might mean complaining but completing the task; a "1" might mean refusal.

These small, family-based rubrics help children connect behavior with outcomes. They learn consistency and quality matter in everything from folding laundry to finishing a project at work. The lesson is simple: good work anywhere prepares you to do good work everywhere.

You can even model this as a parent. Share a story from your job about a time you missed a detail or had to redo something. Maybe you forgot to attach a file, misread a deadline, or sent an email too quickly. Explain how you used feedback to improve. Say, "I didn't get it right the first time either, but I learned to slow down and check my work." Just one story can teach more about quality than a dozen lectures.

As the leader of your household, you can use AI as your assistant in this process. Before asking for help, gather the key details: your child's grade, the subject, and the skills they need to strengthen. Then, prompt AI to help you design a clear, age-appropriate rubric. For example:

"Create a sixth-grade writing rubric with 0-4 scoring levels emphasizing grammar, organization, and clarity, written in student-friendly language."

Or

"Design a science project rubric that highlights use of evidence, creativity, and correct terminology."

By crafting rubrics together, you teach more than writing or math - you model collaboration and problem-solving. You show your child it's normal to ask for help, to clarify expectations, and to take initiative. You are using AI not to replace thinking but to reinforce it.

Once your rubric is created, make it a part of your family's learning rhythm. Review it before assignments begin, not just after grades arrive. Have your child self-assess before submitting work. Ask, "Where do you think this fits on the rubric?" Encourage them to see quality as a process, not a surprise. When feedback comes from the teacher, compare it to your family rubric. Discuss what went well and what could improve.

When emotions flare and they feel defensive, frustrated, or misunderstood, lean on empathy. Say, "I know it's tough to hear this. I've been there too." Remind them feedback isn't failure; it's guidance. Accountability with empathy is your leadership superpower. You are not judging them; you are coaching them. You are helping them build the internal compass that points toward excellence long after school ends.

As your family begins to reason with rubrics, you may notice something powerful happening. Conversations about quality start to feel calmer and clearer. Instead of saying, "Why is this so sloppy?" you can say, "Let's look at where this fits on our scale." Instead of your child saying, "You're never satisfied," they begin to ask, "How can I move from a 2 to a 3?" Rubrics shift the focus from judgment to growth. They make learning measurable and progress visible.

In truth, every family already lives by rubrics. We measure the day by how smoothly it ran, the meal by how it tasted, the week by how balanced it felt. The difference now is you are becoming intentional about it. You are making quality visible. You are teaching your child to see excellence is not perfection, it is attention, consistency, and care.

Rubrics give us language for what we already value. They turn "try harder" into "let's define what better looks like." They turn conflict into collaboration. They remind us , whether in school, work, or family life, we all do our best when we know what the best looks like.

Prompt 35 - Parent: Rubrics and Accountability

Actor: You are an Education Consultant.

Background: My child is in the [enter here] grade and struggles with writing assignments.

Call to Action: Help me design a rubric with 0-4 scoring emphasizing spelling, grammar, organization, and clarity.

Include student-friendly language so they can use it to self-assess.

Data Check: If you need more information, ask me questions before proceeding.

Prompt 36 - Parent: Managing Quality with Rubrics

Actor: You are a Learning Coach

Background: I want to help my [enter here] grader improve the quality of their science homework.

Call to Action: Create a rubric template with clear criteria for use of evidence, clarity of explanations, and correct terminology. Make it easy for us to review their work together and discuss strengths and areas for improvement.

Data Check: If you need more information, ask me questions before proceeding.

Call to Actions for Parents

1. *"Help me explain to my child why quality matters in everyday life."*

2. *"Give me ideas to talk about mistakes as learning opportunities, not failures."*

3. *"Suggest simple ways to make rubric discussions collaborative, not confrontational."*

4. *"Offer strategies for holding my child accountable with empathy."*

5. *"Help me model a growth mindset while reviewing my child's work."*

6. *"List conversation starters to set clear expectations about schoolwork quality."*

7. *"Suggest ways to make checking work a regular family habit."*

8. *"Give me a script to discuss grades and feedback without shaming my child."*

9. *"Help me create a plan to support my child if they continue to struggle."*

10. *"Offer calming phrases I can use when homework conflicts get heated."*

AI prompts for Young Leaders-In-Training

Prompt 37 - Student: Process over Perfection

Actor: You are a Writing Coach

Background: I'm a 6th grader working on a writing assignment for school.

Call to Action: Suggest strategies to help me check my work for spelling, grammar, and organization. Give me simple, kid-friendly feedback I can understand and use on my own.

Data Check: If you need more information, ask me questions before proceeding.

Prompt 38 - Student: Math Review Support

Actor: You are a Homework Helper

Background: I have math homework and I want to make sure my answers are clear and correct. I am [enter here] years old and I am in the [enter here] grade.

Call to Action: Guide me through reviewing my work step by step. Make it easy to understand.

Data Check: If you need more information, ask me questions before proceeding.

Prompt 39 - Student: Time Management

Actor: You are a Study Planner

Background: I often rush through my homework and make mistakes. I am [enter here] years old and I am in the [enter here] grade.

Call to Action: Suggest strategies to help me plan my time better so I can produce quality work. Offer a simple plan or checklist I can follow every time I have homework.

Data Check: If you need more information, ask me questions before proceeding.

CHAPTER 8

Summer Slide

"Our family looks forward to the summertime to get a brief reprieve from the tight schedules required to keep up with school, sports, and work during the school year. What can I do to cultivate my children's critical thinking skills over the summer without over-scheduling my family?"

Every parent feels that sigh of relief when the last bell rings. The backpacks hit the floor, the alarms go silent, and the family calendar breathes for the first time in months. Summer feels like the reward for making it through the long race of the school year. The pace slows. Mornings are gentler. Evenings stretch. Yet, beneath the joy of this pause lingers an important question: How do I keep my child's mind growing when school is out? How do I make space for rest without allowing their curiosity to fall asleep too?

For decades, educators have warned about the *summer slide* that shows a loss of academic skills when children disengage from structured learning. But the problem goes deeper than math facts or reading levels. When children learn to believe growth only happens when school is in session, they internalize a limited view of learning. They begin to see curiosity as something assigned to them instead of something they own. They start to treat knowledge as a requirement instead of a lifelong gift.

This mindset has long-term consequences. Adults who grow up believing learning only happens in classrooms often struggle to adapt later in life. They are more likely to resist change, avoid feedback, and feel overwhelmed by new technologies or professional challenges. They see growth as something that happens *to* them rather than something they can initiate. On the other hand, adults who grew up in homes where learning was part of daily life, through cooking, travel, conversation, exploration, or even failure, carry with them a confidence they can always figure things out. They become self-directed learners, innovators, and problem solvers. Such a mindset begins at home, and it begins with leadership.

As a parent leader, your goal is not to recreate school in your living room. It's to redefine what learning looks like when the textbooks close. Summer is not a break from learning, it is a shift in how learning happens. It's a chance to teach your child growth does not follow a calendar. Their brain does not go on vacation in June. In fact, this is the season when creativity has room to breathe, when curiosity can stretch without grades or deadlines attached.

This one starts with an emotional tone. Children take their cues from you. If you approach summer learning as another obligation, they'll resist. But if you bring a spirit of curiosity, asking questions, exploring ideas, and letting go of perfection, they'll follow your lead. The goal is not to fill every hour with structure, but to fill every experience with purpose. Cooking dinner can become a lesson in fractions, chemistry, and cultural studies. Planting herbs becomes an exploration of plant life (i.e. science) and patience. A day trip to the beach can include journaling observations about tides or marine life. Even cleaning the garage can teach sorting, spatial reasoning, and the satisfaction of completion. Leadership in learning is not about worksheets; it's about awareness.

When you intentionally point out the learning that already exists around your family, you teach your children discovery is everywhere. Watching a movie becomes an exercise in analyzing plot and character motivation. Building a fort turns into a physics experiment about balance and design. Comparing prices at the store becomes a lesson in math, budgeting, and decision-making. The more you label these everyday actions as learning, the more your children begin to internalize that education is a lifestyle, not a season.

Think about the long-term picture. Children who experience learning as a year-round process develop stronger executive function and problem-solving skills. They learn to connect what they study to the real world. They understand curiosity is useful, it can help them plan a trip, fix a mistake, or build something from scratch. Most importantly, they develop the resilience to keep learning even when school ends, teachers change, or life gets difficult. The heart of lifelong leadership.

Setting expectations is crucial. Leadership requires clarity. Don't spring summer learning on your children like an ambush. Invite them into the planning process. Say, "We're going to keep our minds active this summer so that school feels easier in the fall. What kinds of projects or topics sound interesting to you?" When children help design the plan, they take ownership. Developing a sense of autonomy is a powerful motivation.

You can also use technology to keep things organized and collaborative. Use shared family calendars to block out short learning times alongside pool days, picnics, and vacations. Let your child see learning fits naturally within the rhythm of family life. AI can help here too. Use it to create age-appropriate project ideas, build a "summer curiosity challenge," or design family learning contracts in simple, friendly language. When learning is visible, it feels attainable.

Leadership is not about knowing all the answers, it's about modeling how to find them. If your child asks a question and you don't know the answer, resist the urge to bluff. Instead, say, "Excellent question! Let's find out together." Look it up, ask an AI tool, or explore a book. You are showing your child how to reason, research, and verify. Again, you are modeling humility, resourcefulness, and intellectual courage. You're teaching smart people don't always know everything, they keep learning everything.

Of course, resistance will come. At some point, you'll hear, "None of my friends have to do this!" That's your cue to lead with empathy and conviction. Acknowledge their feelings, "I know it feels unfair", then you reframe the comment: "Every family has its own priorities. In ours, we have fun *and* we keep learning." You're not comparing; you're clarifying your family's culture. Great leaders do the same, they define values and live by them, even when others do things differently.

AI can help make those lessons engaging. Craft AI prompts to generate creative family challenges with Call to Actions:

- Turn grocery shopping into a math game about budgeting and nutrition.

- Create trivia questions about the places we'll visit on vacation.

- Design a scavenger hunt that teaches teamwork and observation.

By making AI part of the adventure, you show your children technology can enhance curiosity instead of replacing it. Accountability still matters, but it should be wrapped in compassion. Set goals that are achievable like researching one new question each week or creating a short presentation to share what they've learned. Then, talk about it. Ask, "What was fun about the activity? What was tricky? What do you want to learn next?" Reflection builds metacognition, the ability to think about how we think. It's one of the strongest predictors of lifelong learning success.

If the child forgets or resists, respond as a coach, not a critic. Ask, "What got in the way?" or "What would help you stay on track next time?" A simple shift from blame to problem-solving models emotional intelligence and self-accountability. It shows them progress is not about perfection, it's about persistence.

Summer learning also gives you the opportunity to nurture your child's vision for the future. Ask questions to stretch their imagination: "What do you want to explore? What do you want to create? What kind of impact do you want to have?" Then, use AI tools to help them research those dreams. If your child says he/she wants to be an architect, explore building design together. If they're interested in marine biology, pull up ocean cams and read about coral reefs. If they love gaming, research the science behind how games are built. When they see their interests have real-world pathways, motivation skyrockets. You are connecting present curiosity to future purpose.

Now, zoom out to the larger lesson: Schools are partners in learning, but parents are the architects of it. Schools provide the structure, but homes provide the culture. When families treat learning as something that happens only between September and June, children grow up believing growth is external, something handed to them by teachers. But when families treat learning as an ongoing part of life, children learn growth is internal and something they can lead themselves. They realize curiosity doesn't need a grade to matter. It's the foundation of every meaningful accomplishment they'll ever achieve.

If this sounds like a lot, breathe. You don't need formal lesson plans or color-coded binders. Even ten or twenty minutes of intentional learning a few times a week can make a profound difference. On the days when nothing goes as planned, when everyone's tired or cranky, give yourself grace. Leadership isn't about doing it perfectly; it's about returning to purpose after you fall short. Tell your kids, "I didn't plan anything today, but let's find something new to learn tomorrow." Honesty is leadership in action.

Summer will always be a time for rest, play, and renewal. But as a parent leader, you can expand what it means. You can teach your children curiosity is not confined to classrooms. You can teach them that learning is not punishment, it is a privilege; and growth doesn't pause when school does. Every sunset walk, every question asked, every problem solved is a chance to remind them their mind is a lifelong gift meant to be used, nurtured, and enjoyed.

Your leadership sets the tone. When you treat learning as an everyday habit instead of a seasonal task, you are raising thinkers who know how to lead themselves. You are raising young leaders who see the world as a classroom and life as their curriculum. The most powerful lessons they'll ever learn won't come from a school building. Those lessons will come from watching you, curious, consistent, and committed to growing right alongside them. You will model the kind of leadership that endures long after summer fades.

AI Prompts for Parents

Prompt 40 - Parent: Summer Learning Plan

Actor: You are an Education Consultant

Background: I want to prevent summer slide without over-scheduling my kids.

Call to Action: Suggest weekly topics with critical thinking questions. Make them age-appropriate for 8–12-year-olds and easy to do at home.

Data Check: If you need more information, ask me questions before proceeding.

Prompt 41 - Parent: Empathetic Boundary Setting

Actor: You are a Family Learning Coach

Background: My child complains none of their friends have to do this summer learning.

Call to Action: Help me respond empathetically while setting clear expectations. Provide friendly, practical language I can use in conversation.

Data Check: If you need more information, ask me questions before proceeding.

Prompt 42 - Parent: Empathetic Accountability

Actor: You are a Parent Support Assistant

Background: I want to keep my kids accountable for summer learning without nagging.

Call to Action: Suggest phrases that encourage improvement kindly. Offer language that feels supportive, not critical.

Data Check: If you need more information, ask me questions before proceeding.

AI Prompts for Young Leaders-In-Training

Prompt 43 - Student: Critical Thinking

Actor: You are a Research Assistant

Background: I'm exploring traffic patterns this week. I am [enter here] years old and I am in the [enter here] grade.

Call to Action: Give me questions to ask when observing intersections. Make them simple and clear so I can use them on my own.

Data Check: If you need more information, ask me questions before proceeding.

Prompt 44 - Student: Emotional Intelligence

Actor: You are a Study Helper

Background: I want to do a mini-report on harmony in my family. I am [enter here] years old and I am in the [enter here] grade.

Call to Action: Suggest ideas and questions I can use to learn and teach my family about harmony. Make it fun and easy so I can share what I learn with everyone.

Data Check: If you need more information, ask me questions before proceeding.

Prompt 45 - Student: Summer Slide

Actor: You are a Learning Buddy

Background: I don't want to do schoolwork when my friends don't have to. I am [enter here] years old and I am in the [enter here] grade.

Call to Action: Help me understand why summer learning is important. Explain it in a fair and encouraging way.

Data Check: If you need more information, ask me questions before proceeding.

CHAPTER 9

Entrepreneurship and College

"Our family is in the season of preparing for what comes after high school, the time when years of schoolwork, sports, service projects, late nights, and early mornings all build toward a pivotal choice. Should we be encouraging our child to pursue entrepreneurship to avoid staggering student loans?"

W hether your child is planning for college, considering trade school, dreaming of launching a business, or simply exploring their options, one truth holds steady: Everything you are doing with them now is an investment in their future. As a parent leader, you are the Chief Financial Officer of your household, managing not just money, but mindsets. A strong family leader doesn't just issue orders. They clarify expectations, nurture skills, and prepare their team to make informed, confident decisions. The choice between college and entrepreneurship isn't a competition. Both require resourcefulness, vision, and disciplined planning. The real lesson is teaching your child to think like an entrepreneur about education itself.

When my children were in school, I was a single parent balancing full-time work, home responsibilities, and the hopes of four young dreamers. I didn't have the luxury of guessing how we'd pay for college. I decided early on my children would not start adulthood drowning in debt. I wanted them to have access to the most competitive schools, if it was what they desired, and to get there through effort, not luck. So, I set a standard: in our home, preparation was freedom. They were expected to perform at their best, not because I demanded perfection, but because I refused to let poor preparation close doors they were capable of walking through.

Teaching a child to think like an entrepreneur about education changes everything. Entrepreneurs take initiative, follow through, deliver quality, and seize opportunity. They solve problems instead of waiting for permission. This spirit is what turns an average student into a motivated one. I often told my children their grades, test scores, volunteer work, and leadership roles were part of their "portfolio." Colleges and scholarship committees are investors, and investors look for performance and promise. When students understand this point, their effort takes on new meaning. They stop working just for grades and start working for ownership of their future.

Scholarships are one of the smartest entrepreneurial tools a family can use. They are not favors; they are investments. When a committee awards a scholarship, it is saying, "We believe in your work ethic, your vision, and your ability to deliver." Merit isn't a trait you're born with. It's something you build through diligence, service, and consistency. Scholarships reward the same traits that make successful entrepreneurs: accountability, creativity, and persistence. The earlier your child begins building a track record, the more competitive they become. Every paper, every project, every leadership role is another line in their résumé for future opportunity.

When I wanted my children to take scholarships seriously, I treated the process like a business. The summer before my daughter's senior year, I paid her a wage to spend her days researching and applying for merit-based scholarships. It was her first paid job and one that would pay dividends for a lifetime. Together we created a plan with professional expectations. She defined her value proposition, her academic strengths, her story, her goals and learned how to communicate them authentically and persuasively. We reviewed scholarship opportunities as if they were potential investors. Each essay became a business pitch. She refined her message, tailored it for different audiences, and maintained a portfolio of essays she could adapt and reuse.

We even developed a rubric together to define what "good" looked like. A complete application included clear essays free of errors, well-researched alignments with each organization's mission, and submission before deadlines. Her schedule was flexible, but with structure. She could work wherever she liked, but her chosen work hours had to be honored. She managed her time, communicated when changes were needed, and kept distractions to a minimum. What she was really learning was professional discipline. The same discipline she would need in college and in any entrepreneurial venture.

Some parents might assume their children wouldn't be motivated to follow through, but motivation is built through structure, communication, and visible results. By framing scholarship applications as her summer job, she began to see the direct connection between effort and reward. Each completed application represented a potential return on investment.

Instead of a summer spent earning minimum wage, she was building a six-figure educational portfolio. The experience taught her to see herself as capable of funding her own future. By the end of that summer, she had earned more than $250,000 in scholarship offers. Watching her realize her work had tangible value was one of my proudest moments as a parent.

The lessons she learned mirrored entrepreneurship perfectly. She learned to market herself with honesty and confidence, to research opportunities strategically, and to manage projects independently. She discovered quality and consistency are not accidental - they are habits. She saw that disciplined effort compounds over time, just like financial investments. Those habits followed her into college, through medical school, and beyond. She learned preparation is freedom, and freedom is worth the effort.

AI can be a powerful tool on this journey. It can help your child brainstorm essay topics, refine drafts, and research scholarship opportunities that fit their goals. It can organize application timelines, generate checklists, and simulate interview questions to help them practice their responses. For parents, AI can simplify scheduling, reminders, and organization, reducing the chaos surrounding these high-stakes decisions. The goal is not to have technology do the thinking, but to use it to stay focused on what matters - strategy, quality, and follow-through.

The entrepreneurial mindset extends far beyond scholarship season. It teaches your child how to approach college or business with the same discipline and creativity. A college student with an entrepreneurial mindset doesn't just take classes. They network with professors, seek internships, lead organizations, and build relationships that open future opportunities. They manage their schedules intentionally, balancing rigor with well-being. They see college as a business investment and work to maximize the return.

For children who choose to start a business, the same mindset applies. They must set goals, organize workflows, pitch ideas with clarity, and learn from setbacks without losing focus. Even if one of your children struggles academically, they can still think like an entrepreneur by using the ABCD Framework to design AI prompts to help them build study systems, manage time, and track progress in ways that work for their learning styles.

In the end, the conversation about college versus entrepreneurship is really a conversation about choices. You want your children to make choices today that will create options tomorrow. Poor performance and shortsighted decisions limit freedom. Strong performance and preparation expand it. As their leader, your job is to help them connect daily effort to long-term opportunity. Talk with them about real-world examples in your community, people who built stability through preparation and those whose options were limited by neglecting it. Use those stories to ground the lesson: Freedom isn't found; it's built.

You don't have to do this perfectly. You just have to lead intentionally. Every conversation about goals, effort, and integrity plants seeds of vision. Every scholarship application, every essay, every late night of research is a brick in the foundation of their independence. The message you are sending is simple and profound: Your education is your enterprise. You are the CEO of your own future. You must work with this belief.

When graduation day arrives, you want your child standing on the stage not just with a diploma, but with confidence and clarity about what comes next. Whether they walk into a university, a trade, or a business of their own, they should understand success is not about luck or loans. It's about leadership, stewardship, and the choices made along the way. As the leader of your family, you are shaping not just their next step, but their mindset for a lifetime. You are teaching them preparation creates freedom, discipline builds options, and every choice is a small act of entrepreneurship in the business of life.

AI Prompts for Parents

Prompt 46 - Parent: College Preparation

Actor: You are a College Preparation Consultant.

Background: I want my child to approach high school like an entrepreneur, so they take ownership of competitive performance for scholarships and have choices.

Call to Action: Suggest ways to take an entrepreneurial approach to building a strong academic and extracurricular portfolio. Provide friendly, motivating language I can use in conversation.

Data Check: If you need more information, ask me questions before proceeding.

Prompt 47 - Parent: Emotionally Intelligent Conversations

Actor: You are a Family Strategy Coach

Background: My child doesn't see why working hard matters if they are not even sure about college.

Call to Action: Help me explain the value of their keeping options open. Include examples of college experiences they find attractive.

Data Check: If you need more information, ask me questions before proceeding.

Prompt 48 - Parent: Time Management

Actor: You are an expert on organization and time management.

Background: We are juggling college applications and scholarship deadlines.

Call to Action: Help us create a timeline and checklist. Make it simple enough for my child to take ownership and easy enough for me to manage.

Data Check: If you need more information, ask me questions before proceeding.

AI Prompts for Young Leaders-In-Training

Prompt 49 - Student: Scholarship Preparation

Actor: You are a High School Guidance Counselor.

Background: I want to apply for merit-based scholarships. I excel in [Insert here] and I am interested in these [insert here] topics. I've earned these awards in high school: [insert here].

Call to Action: Help me identify merit-based scholarship opportunities based on my specific achievements and interests.

Data Check: If you need more information, ask me questions before proceeding.

Prompt 50 - Student: Planning for the Future

Actor: You are a High School Guidance Counselor.

Background: I want to be ready for college and maybe start my own business one day.

Call to Action: Suggest actions I can take in high school to prepare for both.

Data Check: If you need more information, ask me questions before proceeding.

Prompt 51 - Student: Planning for the Future

Actor: You are a High School Guidance Counselor.

Background: I don't know what I want to be when I grow up. I like these tasks: [Insert here]. But, I don't like these tasks: [Insert here].

Call to Action: Help me figure out which careers are best suited for me. Are there any free surveys to help me figure out what I want to become?

Data Check: If you need more information, ask me questions before proceeding.

CHAPTER 10

See and Read

I struggled with reading as a kid and don't want my son to face the same challenges. How can I support his early reading skills in simple, manageable ways and fit it into our busy family routine before kindergarten?

When I think about why reading means so much to me, I always go back to my grandmother. She was bedridden with cancer during the last two years of her life, and I spent nearly every moment at her side, talking, playing, and listening to her stories. She told me she didn't hurt as much when I was near her, so naturally, I never wanted to leave. When it came time to start kindergarten, I wailed, loudly and dramatically, until the headmistress called my mother to come take me home. I thought I had won, but my mother had other ideas. She looked me squarely in the eye and said, "If you stay home, you're going to grow up to be an old woman who can't read."

That single statement shaped my life. I already knew my letters, and my grandmother refused to let the challenge stand. From her bed, she tore scraps of paper, wrote words on them, and taught me to sound them out. When I learned enough, she showed me how to fit them together into sentences. Her bed became our classroom, her blanket our chalkboard. She taught me to see reading as a puzzle, a code once cracked, would open a whole new world. Soon I was reading cereal boxes, street signs, and my mother's mail. What started as a challenge became a mission. My grandmother gave me the gift of literacy and with it, the foundation for every success to follow.

Years later, when I became a parent, I taught my children to read using the same linguistic approach my grandmother used, guided by Leonard Bloomfield and Clarence Barnhart's classic book *Let's Read*. The method is beautifully simple: Teach children consistent letter-sound patterns first so they can decode words rather than guess. It starts small, cat, hat, mat, and grows from there. Success builds confidence. Before long, my children were recognizing words at two and reading full books before kindergarten. We didn't have fancy programs or expensive materials. What we had was consistency, curiosity, and connection.

This approach fits perfectly into a modern family routine because it's practical and flexible. You don't need to carve out hours each day. You just need moments of intention. Use AI to help you plan short sessions: "Give me ten simple consonant-vowel-consonant words for my 5-year-old," or "Create three short sentences to practice the letter sound for S." AI can deliver ideas instantly, saving time and giving you structure when your brain is too tired to think.

In our home, reading wasn't a school assignment, it was an adventure. We read everything: books, labels, recipes, street signs, menus, text messages, and even movie captions. We read aloud and listened to each other. No one was ever told "no" when they asked for a new book. We visited libraries, explored used bookstores, and made a family tradition of finding small independent shops on every vacation. My children grew up understanding reading wasn't an obligation; it was a privilege.

Some of our best memories were what we called "reading retreats." We would close the blinds, silence our phones, pile onto my bed, and spend an entire afternoon reading together. If someone found something funny, they'd read it aloud and we'd laugh until our sides hurt. It wasn't school, it was a sanctuary.

Reading was also portable. As a working mother earning my doctorate while raising four children, we spent a lot of time in the car. So, the car became our traveling classroom. One child would read aloud while the others listened, regardless of age. The older ones practiced patience; the younger ones practiced courage. We turned road trips into storytime marathons and often refused to leave the car until the chapter ended. That's leadership at home: Transforming ordinary time into learning time.

Reading teaches patience, focus, and imagination. Technology entertains, but it doesn't replace the mental work of turning words into pictures. Reading strengthens the imagination because it demands participation. When we read *Island of the Blue Dolphins* by Scott O'Dell, a story about a twelve-year-old girl surviving alone on an island, we talked about courage, problem-solving, and faith. My children loved it so much they gifted the book to friends at every birthday party. Find a book like "Island of the Blue Dolphins", a family treasure to enjoy together.

It's easy to let screens take over, but as the leader of your family, you are responsible for protecting your children's attention and imagination. Ask yourself: "Do my children see me reading? Do I give books as gifts? Do I quote favorite lines or read aloud at the dinner table?" Children learn what we model. If we treat reading as essential, they will too. In our home, books were necessities, like food and shelter. In my family, toys were extras, but books were sacred.

Recently, I babysat my two-year-old 'fictive' grandson. I showed him my home library and explained why books matter. A few days after he left, I found his little alphabet book tucked between my journals. He had placed it there on his own, a quiet message the lesson stuck.

Technology can strengthen your leadership, too. On nights when you're too tired to explain why "ph" sounds like "f," create an AI prompt using the ABCD Framework.

Prompt 52 - Parent: Phonics Patterns

Actor: You are a patient reading tutor for a preschooler.

Background: My child is confused by the similar sounds of "ph" and "f."

Call to Action: Explain this pattern in child-friendly language.

Data Check: If you need more information, ask me questions before proceeding.

AI can become your personal literacy assistant and it is always ready whenever you need a spark of creativity or clarity.

You are not just teaching your child to read. You are shaping how they see themselves as learners. Your home is the first classroom, and you are the first teacher. When you pause to sound out a word together, when you ask what they think will happen next, or when you cheer their effort, you are modeling leadership in its purest form: patient, present, and purposeful.

We had nights when we came home late from Tae Kwon Do or an after-school program, and everyone was exhausted. On those nights, I'd say, "Let's just read one page," or "I'll read while you take your bath." A parent leader knows flexibility keeps habits alive. It's not about perfection; it's about presence.

To start, pick one book and read together for five or ten minutes a day. Just short sessions is all it takes. You can find inexpensive copies at thrift stores or local libraries. Make reading part of your daily rhythm, before bed, during breakfast, or in the carpool line. Ask questions, make predictions, and draw scenes from the story. Encourage your children to find letters on packaging while you cook, or words on signs while you drive. Every word they read is another building block of confidence.

AI can help you plan and sustain these experiences. Try prompts like:

- "Give me a list of fun read-aloud books for a 7-year-old."

- "Write five silly tongue twisters using simple consonant sounds."

- "Suggest ways to encourage reading for a busy family that travels often."

Think of it as your literacy strategist, a quiet helper keeping your family learning and connected.

The truth is, many adults still struggle with reading. You may know coworkers, friends, or relatives who avoid reading out loud or mispronounce words. It's not a lack of intelligence, it may be a number of challenges they faced or a lack of early intervention. Poor reading skills can limit confidence, communication, and opportunities. The steps you take now protect your children from struggling later in life.

Your children may not thank you today when you insist on ten minutes of reading. But one day, they will. They'll thank you when they can read anything put in front of them, when they advocate for themselves with confidence, when they dream in words they understand. See this as your return on investment. You are not just the family's Chief Financial Officer, you are the Chief Reading Officer.

Take heart. With consistency, curiosity, and leadership, you can build a family of readers in the midst of a busy, messy, beautiful life. Ten minutes a day is enough. Close the blinds, pile onto the bed, and get lost in books together. Reading is not only a skill, it's an inheritance. You are the leader who passes it on.

AI Prompts for Parents

Prompt 53 Parent: Everyday Reading Strategies

Actor: You are an experienced Family Literacy Coach who understands busy households.

Background: I want to help my child learn to read without adding stress to our already hectic routine. My child is [enter here] years old and I need practical and inexpensive ways to encourage reading.

Call to Action: Suggest practical ways to teach letter sounds and word recognition during everyday family activities.

Data Check: If you need more information, ask me questions before proceeding.

Prompt 54 - Parent: Building Reading Confidence

Actor: You are a Child Development Specialist with expertise in early reading strategies.

Background: My child enjoys stories but resists reading out loud. I want to encourage participation without pressure. My child is [enter here] years old.

Call to Action: Suggest creative strategies to help my child see reading aloud as fun and rewarding.

Data Check: If you need more information, ask me questions before proceeding.

Prompt 55 - Parent: Mobile Reading Routine

Actor: You are an Early Literacy Coach who designs practical routines for busy households.

Background: We spend a lot of time commuting and managing after-school activities. So, it is hard to find a consistent time to sit down and read at home. I want to build my child's confidence as they learn to read, even when they make mistakes. My child is [enter here] years old.

Call to Action: Give me ideas to integrate reading into our commute.

Data Check: If you need more information, ask me questions before proceeding.

AI Prompts for Young Leaders-In-Training

Prompt 56 - Student: Book Selection Support

Actor: You are a Creative Storyteller who loves working with young children.

Background: I want to enjoy reading. I am [enter age here] years old and I am interested in [enter here].

Call to Action: Recommend age appropriate books I might enjoy.

Data Check: If you need more information, ask me questions before proceeding.

Prompt 57 - Student: Daily Vocabulary Practice

Actor: You are a friendly Reading Tutor who explains things clearly for little kids.

Background: I'm learning to read and want to practice a few new words each day. I am [enter here] years old.

Call to Action: Teach me five words I can learn and use in a sentence.

Data Check: If you need more information, ask me questions before proceeding.

Prompt 58 - Student: Travel Reading Games

Actor: You are an Educational Game Designer who knows how to make learning fun.

Background: We spend a lot of time in the car as a family. I am [enter here] years old.

Call to Action: Help me find creative ways to practice reading while I am in the car.

Data Check: If you need more information, ask me questions before proceeding.

CHAPTER 11

When No Means No

"My son does not accept 'no' and will repeatedly nag me when I do not comply with his wishes. What can I do to help him understand when no means no?"

Let me start by saying, you're raising a child with courage. The ability to advocate for oneself is an incredible skill, one that will serve him well in the boardroom, the courtroom, or even at the kitchen table when he's grown. A child who knows how to ask for what he wants is already on his way to becoming a confident adult. But while advocacy is a strength, it becomes a liability when it turns into persistence without boundaries. The goal is not to silence his voice, it's to teach him how and when to use it.

Your job as the CEO of your family enterprise is to channel that strong-willed spirit without surrendering authority. In business, even the most innovative team members know leadership sets the final direction. Your son must learn the same truth at home: A parent leader's "no" is not an opening bid, it's a complete sentence.

Children often test the "no" because they are studying how power and consistency work. They are observing how you make decisions, how you enforce boundaries, and whether your words align with your actions. If you sometimes say "no" when you're frustrated, "yes" when you're tired, or change your answer after a meltdown, your child learns persistence, not respect, and gets results. A single exception can undo weeks of structure. Just as a company loses integrity when it makes exceptions to policy, a family loses authority when a leader's decisions can be negotiated endlessly.

You must teach your child "no" doesn't need a committee vote. It doesn't require a sibling's opinion, a grandparent's override, or a second chance with another parent. The answer remains the same no matter who is asked. However, it does require communication among parents, grandparents, and other stakeholders. There is no need to scan the room for reinforcement or shop for softer authority. When a parent leader has spoken, the discussion is complete. This principle is not about control, it's about security. A child who understands that boundaries are clear and stable learns to trust leadership.

Clarity begins with you. Audit your leadership brand. Are your "no" decisions rooted in love and logic, or are they reactions to fatigue or frustration? Children can tell. A thoughtful "no" delivered calmly builds respect, but an impulsive "no" invites rebellion. A parent leader must know why they are saying no, to protect safety, reinforce values, teach discipline, or preserve balance. When your "no" is intentional, your child feels its purpose, even when she disagrees.

Once your motives are clear, define what's negotiable and what's not. Families thrive on operational clarity. Just like a strong organization, your family should have categories: **negotiable** decisions (like bedtime on weekends, the order of chores, or dessert choices) and **non-negotiable** decisions (like safety, respect, and honesty). When your child knows which is which, you eliminate the gray areas that fuel endless debate. This distinction doesn't suppress independence,it organizes it. It teaches your child the structure of authority and builds trust rather than anger or fear.

Negotiation still has a place in leadership, and your home is the perfect training ground. You want your child to learn how to present a case respectfully, use data, and propose solutions. This skill will serve him in adulthood far beyond the dinner table. But negotiation must have boundaries too. In my home, Saturdays were "Appeals Day." My children could challenge any family rule or request a policy change, but only on Saturdays after lunch. Why? Because weekday decisions were final, and weekday energy was limited. Saturday appeals gave everyone time to think clearly and engage respectfully.

Each child had to write their argument on an index card before the meeting. It wasn't about formality; it was about preparation. A written argument required focus and logic. It turned emotional pleading into structured reasoning. "I want a later bedtime because everyone else stays up later", became "If my bedtime changes from 8:00 p.m. to 8:30 p.m., I will use the extra time to read quietly, which helps me relax and improves my vocabulary." Now, that's a compelling case. Sometimes they won, and the rule was adjusted. Other times, I said, "no" and explained why. Either way, the process was dignified. It taught them to prepare and communicate; it also taught them a leader's "no" may still stand.

Again, you are modeling leadership with compassion and consistency. You are teaching your child "no" is not a personal rejection, it's an act of love. "No" can mean "not now," "not safe," or "not best." Once "no" is spoken, it doesn't need to be repeated, debated, or verified with another adult.

We have all seen adults who crumble when told 'no'. They yell, manipulate, or escalate. They may not have been taught emotional regulation in childhood. Perhaps, they learned persistence beats boundaries. You are breaking the pattern. By holding firm, you are preparing your child to handle disappointment, delay, and discipline, the hallmarks of emotional maturity.

You can even make this structure visible. Ask AI to help you write a simple family policy document defining your negotiable and non-negotiable categories. Review it as a family. Encourage input but maintain leadership. Use the ABCD Framework to get started:

Prompt 59 - Parent: Empathetic Accountability

Actor: You are a Family Coach who helps parents create healthy household rules.

Background: My child struggles to accept "no" and often seeks validation from other adults after I've given a final answer.

Call to Action: Help me draft a short, clear family policy explaining when "no means no" and how appeal sessions will work.

Data Check: If you need more information, ask me questions before proceeding.

Structure protects peace. Knowing "appeals" happen only on Saturdays, for example, frees the week from constant negotiation. You can even keep a family "Appeals Log," where your child writes down the issues they want to revisit. Most will be forgotten by Saturday, but those that matter will come back with thought and preparation.

When "no" is challenged, stay calm. Humor can defuse tension faster than anger. Say, "Good try, but the decision is final. Please add it to your Saturday list of appeals." Smile, and move on. Calm, predictable enforcement builds confidence. A parent who stays centered while holding boundaries sends a powerful message: Rules are not emotional, they are structural.

The greatest leadership lesson you can model is authority doesn't need to shout. When your "no" is grounded in purpose, consistency, and calm, your child learns to respect it without fear. They also learn leadership does not require validation. This lesson will serve them far beyond childhood. When they are older and face peer pressure, they won't look around the room to see what everyone else thinks before deciding. They'll trust their own boundaries. They'll know a firm "no" is not weakness, it's strength.

You are not raising a follower; you are raising a future leader. Leaders who know how to say no without apology are the ones who protect their integrity, time, and energy. They don't crumble under pressure or seek approval for every decision. They listen, they weigh, they decide, and they stand firm. Your consistency at home lays the foundation.

So, when your child pushes back for the fifth time, stay steady. Say, "I've already answered. You may not like it, but the answer stands." If they look to another adult for reinforcement, calmly remind them, "I am your parent. My decision is final." Over time, your tone, not your volume, will teach them calm authority is immovable.

One day, your child will thank you. Not for the times you said 'yes', but for the times you held your ground. For the times you were the leader who didn't flinch when pushed. For the structure you provided and the safety you created, even when they were angry. They'll understand your "no" wasn't rejection, it was leadership in action.

You are the parent leader. Your word carries weight because it is grounded in love, wisdom, and consistency. A leader's "no" is never harsh, it's steady, clear, and complete. A lesson will follow your child everywhere, shaping how they lead, love, and live.

AI Prompts for Parents

Prompt 60 - Parent: Setting Boundaries

Actor: You are a Parenting Coach

Background: My child refuses to accept "no" and constantly negotiates, leading to daily power struggles.

Call to Action: Help me draft clear family rules that separate negotiables from non-negotiables. Include kid-friendly, but firm, language I can use to explain the rules calmly and clearly.

Data Check: If you need more information, ask me questions before proceeding.

Prompt 61 - Parent: Teaching Respectful Communication

Actor: You are a Child Development Specialist

Background: I want to set up a weekly negotiation session with my child to reduce conflict.

Call to Action: Suggest a structure and guidelines for these meetings that teach respectful argument and emotional control for a [insert here] year old child. Make sure it's easy for busy parents to follow and implement.

Data Check: If you need more information, ask me questions before proceeding.

Prompt 62 - Parent: Managing Emotions

Actor: You are a Family Communication Strategist

Background: My child gets upset when I say "no" and struggles to manage disappointment.

Call to Action: Give me phrases to hold firm boundaries while showing empathy. Help me keep my emotional tone calm.

Data Check: If you need more information, ask me questions before proceeding.

AI Prompts for Young Leaders-In-Training

Prompt 63 - Student: Negotiating Family Rules

Actor: You are a Negotiation Coach

Background: I am in the [enter here] grade and I am [enter here] years old. I want to convince my parent(s) to change one of our family's rules.

Call to Action: This is what I want to change [insert here]. Suggest 3 strong, respectful reasons for my request.

Data Check: If you need more information, ask me questions before proceeding.

Prompt 64 - Student: Learning Respectful Communication

Actor: You are an Emotional Intelligence Coach.

Background: I am in the [enter here] grade and I am [enter here] years old. I feel really frustrated and want to yell when my parent(s) says "no."

Call to Action: Teach me ways to calm down and explain what I want without yelling.

Data Check: If you need more information, ask me questions before proceeding.

Prompt 65 - Student: Learning to Negotiate

Actor: You are a Writing Tutor

Background: I am in the [enter here] grade and I am [enter here] years old. I'm supposed to write my appeal for our family negotiation day.

Call to Action: Help me organize my argument clearly and respectfully. Make sure it's short but convincing so my parent(s) will take it seriously.

Data Check: If you need more information, ask me questions before proceeding.

CHAPTER 12

Tears and More Tears

"My child cries easily and it's starting to cause problems at home and school. The teacher even sent a note suggesting a medical evaluation. Is it normal for a third grader to cry this much?"

Welcome to your role as Chief Family Officer of the House of Emotional Expression. Parenting is leadership at its most personal. You are the founder, strategist, and crisis communications director of your family. When your third grader bursts into tears for the third time today, you are not managing a small leak, you are tending to the plumbing system of their emotional development. And yes, crying at this age is normal. Not because your child is behind, but because they are still learning how to be human.

Start with the basics. Just as any responsible leader checks the foundation before rebuilding, schedule a visit with your pediatrician to rule out physical or neurological issues. Sometimes, anxiety, sensory challenges, or language frustrations hide behind excessive tears. If everything checks out, you don't have a problem child, you have an expressive one who needs tools, structure, and your steady guidance. This is where your leadership begins, not with punishment or panic, but with empathy and calm.

Children learn how to manage emotions by watching the emotional tone you set. If you respond to crying with irritation, sarcasm, or dismissal, you teach them feelings are unsafe. It's like running an organization where employees are afraid to share problems until the system collapses. Instead, stay composed and curious. Say, "I see you're upset. Can you tell me what's going on?" or, "Let's take a few deep breaths together." Your calm presence becomes their emotional template. When your child learns big feelings can be met with safety, not shame, you're shaping the culture of your home.

When everyone is calm, ask "what happens right before the tears begin". Are they frustrated because they can't find the right words? Are they overwhelmed by noise or disappointment? Or are they using tears as a way to gain attention or change your mind? Understanding the reason behind the tears is how you lead with clarity instead of confusion.

It's important to explain that emotions are important but not the boss of the family. It's okay to cry when you are sad, hurt, or disappointed, but not every frustration requires tears. You're teaching your child to recognize emotions are information, not instruction. Just as a good leader knows when to speak and when to listen, your child is learning that feelings deserve acknowledgment but not control.

Many tears come from frustration and the inability to express complex emotions. Your role as coach becomes important here. Teach your child words for what they feel, such as disappointed, embarrassed, left out, frustrated, or worried. Then model how to use them. Role-play: "What could you say if your friend took your toy?" or "How could you tell your teacher you feel left out?"

You are training your child to process emotion through language rather than meltdown. This is problem-solving in action.

AI can help you make these lessons engaging. Use your ABCD Framework to create prompts such as:

Prompt 66 Parent: Highly Reactive Emotions

> **Actor:** You are a child development coach.
>
> **Background:** My child cries when frustrated and struggles to express emotions in words.
>
> **Call to Action:** Create games or conversation starters that help children identify and manage their emotions in healthy ways.
>
> **Data Check:** If you need more information, ask me questions before proceeding.

When you hold accountability with empathy, you build trust. Many parents rush to stop crying because they feel uncomfortable or embarrassed; but, tears are not a malfunction. They are the body's built-in stress relief system, filled with the hormones needed to calm the nervous system. Crying helps the body reset, just as rebooting helps a computer run smoothly. It is perfectly normal and even necessary. Children need to hear that crying is okay, free from stereotypes that suggest boys who cry are weak or girls who cry are manipulative. Neither belief serves them. Say to your son, "Crying doesn't make you weak, it shows you care." Say to your daughter, "It's okay to cry, but you must also learn to talk about what you want and how to express it clearly." You are building a home culture where emotion and strength coexist.

Sometimes tears can be used to test boundaries. Children are excellent at reading systems for weakness. If crying gets them what they want, they will keep using it. This is when your leadership must be consistent. Say, "I see you're upset, but crying won't change my answer." A compelling response acknowledging emotion without rewarding manipulation. A parent leader's "no" is a complete sentence, not a debate invitation, not a committee vote. When you hold your ground consistently, you are teaching stability. Your child learns emotions can be felt fully without driving decisions.

If your child's crying begins affecting them at school or fracturing friendships, work collaboratively with their teacher. Approach it as a team: gather data, look for patterns, and discuss strategies. Leadership means responding to information, not reacting to discomfort. If a counselor or psychologist recommends evaluation, accept it as support, not failure. Early guidance can give your child lasting tools for emotional regulation.

Now, look inward. Children learn how to manage emotion by watching you manage yours. If you slam doors or shut down when you're overwhelmed, your child learns emotions are dangerous. Instead, narrate your self-regulation: "I'm really frustrated right now, so I'm going to take a breath before I respond." or "I'm sad today, and that's okay. I'll take a walk, listen to music, and give myself time to feel better." You are showing adults also experience strong feelings and feelings can be managed with grace.

If you find yourself 'triggered' by your child's tears, pause and reflect. Ask, "What about this situation is bothering me?" Sometimes our discomfort is connected to how our own emotions were handled when we were young. AI can help you explore this safely.

Use an AI prompt in the ABCD Framework. See an example below:

Prompt 67 - Parent: Emotional Regulation

Actor: You are a parenting coach.

Background: My child's crying makes me anxious or angry.

Call to Action: Help me explore why I react this way and suggest ways to stay calm and supportive.

Data Check: If you need more information, ask me questions before proceeding.

Every family benefits from calm-down rituals. Businesses use cooling-off periods before major decisions, families can too. When your child is emotional, invite them to sit quietly in a calm space. Say, "You can sit here until your body feels settled. Then we can talk." This teaches composure before communication.

Children who learn to manage big emotions become adults who can face challenges with resilience. A growth mindset is understanding emotion is not the enemy of strength but a part of it. Remind your child, "You're learning to manage your feelings, even adults practice managing emotions." Celebrate progress instead of demanding perfection.

Your goal is not to raise a child who never cries. Your goal is to raise a child who understands why they cry, who can put their feelings into words, and who knows how to recover with confidence. Leadership at home is not about control. It's about teaching composure, compassion, and consistency. The tears may come, but they will not last forever. Each one is a lesson in self-awareness and recovery, and when your child learns to pause, to breathe, and to try again, you'll see it clearly: Your calm fostered their strength. That is leadership in motion.

AI Prompts for Parents

Prompt 68 - Parent: Clarity is Kindness

Actor: You are a Parenting Coach

Background: My child cries easily when they are frustrated, especially with schoolwork or chores.

Call to Action: I want to respond calmly without giving in or shaming them. Give me five kind phrases I can use when my child starts crying out of frustration. But, I want to hold the boundary, too.

Data Check: If you need more information, ask me questions before proceeding.

Prompt 69 - Parent: Emotional Vocabulary Building

Actor: You are an Emotional Intelligence Counselor.

Background: I want to help my child develop a larger emotional vocabulary.

Call to Action: List 10 emotional words with simple kid-friendly definitions *I* can teach my child so they can express themselves.

Data Check: If you need more information, ask me questions before proceeding.

Prompt 70 - Parent: Self-Reflection

Actor: You are a psychologist, specializing in family counseling.

Background: I get triggered when my child cries.

Call to Action: Guide me through three reflection questions to help me understand why my child's crying frustrates me and how I can stay calm.

Data Check: If you need more information, ask me questions before proceeding.

AI Prompts for Young Leaders-In-Training

Prompt 71 - Student: Emotional Resilience Practice

Actor: You are an Emotional Coach

Background: I get really upset and cry when I lose games or when things are too hard. I am [enter here] years old and I am in the [enter here] grade.

Call to Action: Help me figure out three things I can do when I feel like crying because I'm frustrated.

Data Check: If you need more information, ask me questions before proceeding.

Prompt 72 - Student: Emotional Intelligence

Actor: You are a Communication Specialist

Background: I want to tell people how I feel without crying so much. I am [enter here] years old and I am in the [enter here] grade.

Call to Action: Give me three sentences I can use to tell someone I'm sad or mad without crying.

Data Check: If you need more information, ask me questions before proceeding.

Prompt 73 - Student: Emotional Regulation

Actor: You are a Problem-Solving Coach

Background: I want to make a plan for what I can do next time I feel like crying. I am [enter here] years old and I am in the [enter here] grade.

Call to Action: Help me come up with three steps I can follow to calm down when I feel like crying.

Data Check: If you need more information, ask me questions before proceeding.

CHAPTER 13

Getting on Blue

"This is the fifth day in a row my son, Brandon, has gotten on 'blue' for talking. I've taken all his toys, scolded, and punished him, but he keeps getting in trouble for talking. What can I do?"

First, take a deep breath and recognize what is really happening. You are raising a communicator, a child who is curious, expressive and confident enough to use his voice. There are far worse notes to receive from a teacher than he "talks too much." You are not hearing "threw a chair" or "skipped class." You are hearing about a young leader in training who has not yet learned when to deliver the message and when to wait. Your job as the parent leader is not to silence him, but to help him direct his energy.

Start by reframing the behavior before you rush into discipline. Talking in class is not rebellion or defiance. It is an unshaped strength. Brandon's enthusiasm, curiosity, and love for connection are the same qualities it will take for him to become a persuasive communicator, negotiator, or leader in the future. Your challenge is to help him transform impulsive chatter into intentional communication.

Every leadership journey begins with emotional intelligence. Your emotional tone sets the temperature of your home. If every "blue day" becomes a storm of punishment, Brandon will begin to associate his natural gift of expression with shame. That is like telling your company's marketing department to stop sharing ideas and then wondering why creativity disappears. Instead, hold steady and remember your goal. You want to send the message, "Your voice is valuable, but you must learn how to use it well."

Now, take a leadership audit. You have scolded, taken away toys, and punished repeatedly, yet the behavior continues. This is data. It means your strategy is not addressing the real issue. Before you decide on another consequence, gather information. Ask yourself: Is my child talking because he is bored? Is he confused about the lesson? Is he processing information out loud? Is he seeking attention or reassurance? Each reason calls for a different type of coaching.

If he is talking because he needs help understanding instructions, then he is demonstrating resourcefulness, not rebellion. Praise the instinct, and then teach him how to seek help in a respectful way. If the talking is impulsive, remember impulse control is a developmental skill and it takes time to develop it. It is not a sign of disrespect. Guide him toward regulating impulses without erasing his natural enthusiasm.

Help him channel his talking productively. Explain why class time is for listening and learning, and there will be appropriate times for sharing ideas. Work with him to identify those moments. "When your teacher is giving directions, it is time to listen. When she calls for volunteers, it is your time to share." This is how you teach the rhythm of communication.

AI POWERED LEADERSHIP FOR MODERN FAMILIES

You can create a home-based training ground for these lessons. Sit with Brandon and co-create a set of "talking rules." Keep them positive and clear:

- We listen quietly when others are speaking.

- We raise our hands when we want to share.

- We write down thoughts and ideas, if we cannot say them right away.

Invite Brandon to decorate the list and hang it somewhere visible. When children help design the rules, they feel responsible for following them. This is leadership psychology in action.

You can turn this exercise into a small leadership project using AI as your creative assistant. Remember, AI is a tool, not a shortcut. Many parents skip reflection by typing quick prompts and accepting whatever appears. The ABCD Framework exists to help you stay thoughtful and strategic.

Here is how to use it:

Prompt 74 - Parent: Impulse Control Plan

Actor: You are a child behavior specialist.

Background: My 8-year-old talks excessively in class and struggles with impulse control. I want to teach him respectful communication and self-regulation.

Call to Action: Create a communication plan with age-appropriate strategies for when to speak, how to wait, and how to track improvement.

Data Check: If you need more information, ask me questions before proceeding.

This process forces you to think deeply about your child's needs before asking AI to respond. It keeps the human heart of parenting intact while using technology for structure and support.

Model the patience you want Brandon to learn. When he interrupts, say calmly, "I hear you, but it is my turn to speak. I will listen when I am finished." This demonstrates respect and control. Children learn emotional regulation through observation. They watch how you handle frustration more than they listen to what you say about it.

You can build his impulse control through playful practice. Try "The Talking Game," where everyone waits three seconds before responding in a conversation. Or bring back "Mother, May I?" where each move requires patience and permission. Have family "quiet rounds" where you communicate with gestures or notes. Make it lighthearted, not disciplinary. These moments teach emotional stamina, timing, and self-restraint.

If the talking continues, use consequences that connect directly to the behavior. For instance, if he interrupts during quiet work, he may lose speaking privileges for a short period. If he follows the rules, reward him with extra time to share stories later in the day. The consequence should always teach, not just punish. It reinforces the connection between actions and outcomes.

AI can help you brainstorm creative consequences, but resist the urge to type, "What punishment works for a talkative child?" Instead, ask, "Help me design natural, respectful consequences that teach responsibility while maintaining my child's confidence." The wording matters. It keeps your focus on growth instead of control.

Your approach should blend empathy and accountability. If Brandon breaks a rule, say, "I know you wanted to share that story, but it was not the right time. You can tell me later." This preserves his dignity while teaching boundaries. Accountability delivered with calm energy builds trust.

Now, nurture a growth mindset. The goal is not simply to make Brandon quiet; it is to help him grow into a self-aware communicator. Explain the why. "Your words matter. The secret is learning when your words build connections and when they cause distraction." Then, celebrate progress. "I noticed you waited your turn to speak today. That shows maturity. I am proud of you." Encouragement motivates change more effectively than criticism.

If the issue continues, collaborate with his teacher. Ask about classroom dynamics, seating arrangements, or the lesson structure. Teachers usually welcome parents who show up as partners instead of critics. If you observe that Brandon is being singled out, coach him on maintaining composure and confidence. Tell him, "You cannot control how others respond to you, but you can always control how you respond to them." Teach him that grace under pressure is a superpower.

Turn communication practice into a family routine. Have family dinners where each person gets a turn to talk while others listen. Use a talking stick, a method that indigenous peoples of North America have used for generations to promote fairness and respect in dialogue. This tradition teaches patience, inclusion, and thoughtful communication.

If nothing else works, take leadership action by showing up. With the teacher's permission, sit in class quietly for a day. Not as punishment, but as partnership. Your calm presence communicates that the family's values travel with your child. You are showing him you are invested in his growth and willing

119

to walk alongside him until he masters self-regulation. Once he demonstrates progress, you can step back.

Above all, remember the long game. You are not managing a noisy child. You are raising a communicator who will one day persuade, teach, and lead others. Your goal is not silence but skill. When you stay patient, consistent, and kind, you are teaching him how to manage his energy, not suppress it.

Leadership at home requires patience, strategy, and reflection. By combining emotional intelligence, clear expectations, empathetic accountability, and calm modeling, you are teaching Brandon the art of communication. When you use AI thoughtfully through your ABCD Framework, you remind yourself technology is a tool to guide reflection, not replace wisdom.

You are building a home culture of respect, awareness, and discipline, one conversation at a time. Years from now, when Brandon stands before an audience with a microphone in his hand, speaking with clarity and confidence, you will remember these early "blue days" with a smile. You did not stop him from talking. You taught him how to use his voice well.

AI Prompts for Parents

Prompt 75 - Parent: Respectful Communication

Actor: You are a Parenting Coach.

Background: My child keeps getting in trouble for talking too much at school. He/she is in the [enter here] grade and is [enter here] years old.

Call to Action: I want to set clear family expectations about when talking is appropriate. Provide 10 kid-friendly rules about talking impulsively I can explain easily.

Data Check: If you need more information, ask me questions before proceeding.

Prompt 76 - Parent: Positive Direction

Actor: You are a Child Behavior Specialist

Background: I want to address my child's talking without shaming them or crushing their confidence. He/she is in the [enter here] grade and is [enter here] years old.

Call to Action: Provide me with 5 phrases to use when my child talks at the wrong time so I can redirect them kindly.

Data Check: If you need more information, ask me questions before proceeding.

Prompt 77 - Parent: Parent-Teacher Partnership

Actor: You are a Family Systems Therapist

Background: I want to partner with my child's teacher to solve the problem. He/she is in the [enter here] grade and is [enter here] years old.

Call to Action: Help me write a respectful email to my child's teacher asking if I can sit in on one of his classes.

Data Check: If you need more information, ask me questions before proceeding.

AI Prompts for Young Leaders-in-Training

Prompt 78- Student: Practicing Self-Control

Actor: You are a Communication Coach

Background: I get in trouble for talking during class. I am in the [enter here] grade and I am [enter here] years old.

Call to Action: Suggest 3 things I can do instead of talking when I have something to say in class.

Data Check: If you need more information, ask me questions before proceeding.

Prompt 79 - Student: Self-Management

Actor: You are an Emotional Intelligence Counselor

Background: I feel like I can't stop talking even when I know I should. I am in the [enter here] grade and I am [enter here] years old.

Call to Action: I want help controlling the urge to talk.

Data Check: If you need more information, ask me questions before proceeding.

Prompt 80 - Student: Respectful Inquiry

Actor: You are a Writing Coach

Background: I want to remember my ideas for later instead of interrupting. I am in the [enter here] grade and I am [enter here] years old.

Call to Action: Help me identify ways to keep up with my ideas and questions.

Data Check: If you need more information, ask me questions before proceeding.

CHAPTER 14

My 16-Year-Old Just Called Me a B&^%*

"It started like a hundred other nights, the kitchen looked like a cooking show explosion, and you asked (for what felt like the hundredth time) for your teen to clean it up. From across the counter, the child, the same one you taught to say 'please' and 'thank you', looks you dead in the eye and called you a b&^%."*

L et us start with the gut punch. You hear it. The word you never imagined your own child would say. For a split second, the air leaves your body. Maybe you freeze. Maybe your blood boils. After all the sacrifices you have made for your child, being called 'the b-word' by your beloved teenager can feel like a personal earthquake. Fury, sadness, and disbelief rise all at once. You may even question how you raised this person who just hurled the unthinkable at you. Whatever your reaction, you are not alone. Every parent eventually faces a version of this moment when the child they love says something shocking, cruel, or deeply disrespectful.

Now, let us be clear. Your child's words were unacceptable. But, this moment does not define your entire relationship. It is one data point in a much longer story. It tells you something about where your teen is emotionally, and possibly where you are, too. The _goal_ is not to shame or destroy trust but to understand and lead.

Your teenager has just tested one of the sharpest boundaries between adolescence and adulthood, respect. In their world, that word might have come out of frustration, humiliation, or a desperate attempt to seize control. To them, it may have felt justified. To you, it felt like betrayal. Both feelings are real, but one of you must lead.

You are the thermostat, not the thermometer. You set the emotional tone. Meaning when your child turns up the heat, you must remain steady. Take a pause. Walk away if needed. Drink some water. Count to fifty. Pray. Remind yourself leadership in the home means being calm in the storm.

When teenagers lash out, it is rarely about the exact word they choose. It is usually about power, pain, or fear. Teenagers often feel cornered or unheard. They use extreme words because they lack better tools. This does not excuse the disrespect, but it helps decode it.

This is where your emotional intelligence matters most. Do not meet your child's heat with your own. Take time to process before you respond. Think of yourself as an investigator rather than a prosecutor. Ask yourself: "What was happening before those words came out? Did I say no to something? Did my child feel embarrassed, powerless, or compared to someone else?" The goal is not to justify what happened, but to understand the emotional pattern that triggered it.

Timing is crucial. You cannot lead a conversation in the middle of a storm. Wait until both of you have cooled off. When you do speak, begin with the emotion, not the insult. Say, "I could tell you were angry earlier. I was deeply hurt and disappointed by what you said. Let's talk about what was really going on." A single, vulnerable, statement accomplishes two leadership goals. It holds a clear boundary and extends empathy.

If your teenager remains defensive, that is okay. You are planting a seed. The important thing is that you remain composed. Even if they roll their eyes, your steady tone teaches them what maturity looks like under pressure.

Disrespect requires accountability, not humiliation. Grounding your child in front of siblings or launching into a long lecture about how you were raised is not leadership, it is theater. Quiet consequences are far more effective and preserve dignity on both sides. A calm, private discussion might sound like this: "You do not have to like every rule, but calling me names crosses a line. We are going to take a break from privileges until you show emotional regulation."

Emotional regulation should become part of your family vocabulary. Teens crave independence, but they must also know love does not vanish when boundaries are enforced. Consequences are not about punishment; they are about teaching responsibility and self-awareness. That is what leadership looks like in a family.

Before you re-engage with your teen, take time to examine your own emotions. Ask yourself, "what this moment stirred in you? Did it tap into an old wound from your own childhood or from a past relationship?" Sometimes, a parent's reaction is larger than the event because it awakens something unresolved within you.

When it happens, be honest with yourself. The pain may not be only about this moment. It may be about the times you were disrespected or unheard in your own past. If so, this is an opportunity for healing, not projection. Pause, breathe, and choose a response that reflects your values rather than your hurt. Leadership requires self-regulation before correction.

Now, let us address how to repair the relationship after the explosion. When the air clears, schedule what I call a "repair conversation." This is not a lecture or interrogation. It is a guided conversation meant to restore connection and rebuild trust. You might begin with, "I want to talk about what happened earlier. The words you used hurt me. I would like to understand what was behind the anger." Keep the focus on emotion and behavior, not character. You are addressing the action without labeling the person.

Teens need to know two things after a conflict: Your love is unshaken, and your boundaries remain firm. When you demonstrate calm accountability, you teach them relationships can recover without shame.

This is an excellent time to use AI thoughtfully. Write an AI prompt using the ABCD Framework to prepare for your repair conversation. Do not take shortcuts by using the call to action without the complete ABCD Framework, such as, "how do I make my teen apologize?" Instead, use intention and include your Call to Action in the ABCD Framework:

Prompt 81 - Parent: Family Repair Dialogue

Actor: You are a family therapist who specializes in adolescent communication.

Background: My 16-year-old called me a disrespectful name during an argument about chores. I want to have a repair conversation focused on teaching empathy and accountability while keeping the tone calm and constructive.

Call to Action: Write a script I can adapt for a family meeting that models emotional maturity and clear boundaries.

Data Check: If you need more information, ask me questions before proceeding.

This type of prompting helps you reflect before acting. It also teaches your child technology can serve growth when used with thoughtfulness. The ABCD Framework keeps you from reacting impulsively and reminds you parenting with AI should be reflective, not reactive.

When you are ready to talk again, shift from correction to collaboration. Say, "That word hurt me. I know you were angry, but we cannot talk to each other in that way. Tell me what was behind the anger." Then, guide your child toward accountability. "How do you think we can handle anger differently next time?" You are transforming the incident into a leadership exercise in emotional intelligence.

Your teenager's apology, if it comes, may be slow or reluctant. Accept it with grace. If it does not come at all, show forgiveness anyway, to model emotional maturity. Teenagers may not remember every rule you enforced, but they will remember how you handled their lowest moment. They will remember whether you met their rage with rage or with restraint.

When the moment feels raw and unbearable, remember this: You are an adult. You are the standard. Say, "Time-out. You crossed a line. We will talk later." Then step away and calm yourself. Go to a quiet place, grab your phone, and use AI to rehearse your next steps. Ask AI for strategies that help you manage your tone, your breathing, and your words. Technology can be a mirror that helps you regulate before re-engaging.

Once you are both ready, discuss what repair looks like. It may involve an apology, an act of service, or a written reflection. The goal is to teach responsibility without breaking the connection. Accountability does not have to sound harsh. It can sound like the need for reflection.

Try this: "When you called me that name, it made me feel disrespected and hurt. It also made it harder for me to hear what you were trying to say. How can we make sure this does not happen again?" This shifts the focus from punishment to partnership.

You are also modeling problem-solving, one of the core leadership principles. Teenagers must learn that relationships require effort and recovery. When they see you handle conflict with balance, they learn that love and limits can coexist.

Over time, your calm consistency will become the quiet teacher. Even if your teen does not acknowledge it now, the lesson will stay. One day, when they are grown, they will remember you did not crumble or retaliate. They will remember you led with love.

Parenting teenagers means navigating storms without losing your center. It means remembering your child's disrespect is not a final verdict; it is an emotional flare that says, "I am overwhelmed and out of tools." Your response can either escalate the fire or teach a new language of emotional regulation.

This is how you raise emotionally intelligent adults. Not by avoiding conflict, but by modeling how to move through it with grace. You are teaching your child leadership does not mean domination or perfection. It means recovery, reflection, and consistent growth.

I know how hard this is. I have been there. I raised my voice when I should have paused. I reacted when I should have reflected. And I regretted it later. You can still choose grace in the moment. You can return disrespect with calm strength. That is a mic-drop moment of leadership.

Here is what may comfort you most. This too will become a story. One day, your 16-year-old will be 36, standing in her own kitchen, saying to her own teenager, "Do not talk to me like that!" She will pause because she will remember how you handled her worst day. That memory will shape how she leads her own family. That is your legacy - leadership and love.

No matter how imperfectly we parent, our children have free will. They will make unwise choices. They will say things they do not mean. Our job is to heal our own wounds, do our best, and trust that consistency will one day be seen for what it truly was: Love expressed through boundaries, patience, and courage.

If you are reading this after a long night of tears and regret, please know you are not alone. Every difficult moment is another chance to lead. Your love, even when tested, is still the greatest teacher your child will ever have.

Here's how to use the "ABCD Framework" to guide your next move with AI when emotions run high:

Prompt 82 - Parent: Calm Responses

Actor: You are a Family Therapist who helps parents respond calmly when teens use disrespectful language.

Background: My 16-year-old just called me a [insert here] after I set a boundary. I'm hurt and angry but want to handle this wisely.

Call to Action: Help me find language to respond calmly while still addressing the disrespect. And more importantly, help me keep from holding this against my child.

Data Check: If you need more information, ask me questions before proceeding.

Prompt 83 - Parent: Emotional Recovery Guidance

Actor: You are a Parenting Coach who helps write calm conversation scripts.

Background: My 16-year-old called me a [insert word here] during an argument. And it shattered something in me. I love my daughter more than I love myself. Never in a thousand years did I think someone that I love so completely would call me such a name.

Call to Action: Help me stop crying, avoid shutting down, and put myself back together. How can I separate the behavior from my child?

Data Check: If you need more information, ask me questions before proceeding.

Prompt 84 - Parent: Empathetic Response Guidance

Actor: You are a Parenting Coach who helps write calm conversation scripts.

Background: My 16-year-old called me a [insert word here] during an argument. I want to address it without yelling or shutting down.

Call to Action: Provide me with a calm script that expresses how I feel and demonstrates empathy.

Data Check: If you need more information, ask me questions before proceeding.

| AI Prompts for Young Leaders-in-Training

Prompt 85 - Student: Apology Reflection

Actor: You are a Teen Reflection Coach.

Background: I called my [Mom/Dad/Sibling] a hurtful name during an argument. I was angry and frustrated, but now I feel guilty.

Call to Action: Help me understand why I said it, what I was really feeling, and how I can apologize sincerely.

Data Check: If you need more information, ask me questions before proceeding.

Prompt 86 - Student: Calm Response Guidance

Actor: You are an Emotional Intelligence Mentor.

Background: I lose control and say things I regret when I am angry with my parents.

Call to Action: Teach me how to pause, breathe, and respond calmly instead of shouting or cursing.

Data Check: If you need more information, ask me questions before proceeding.

Prompt 87 - Student: Anger Awareness Plan

Actor: You are a Youth Therapist who helps teens handle anger in healthy ways.

Background: I often say mean things when I feel cornered or disrespected. I do not want to keep reacting or repeating this behavior.

Call to Action: Help me create a plan to recognize my triggers and express anger without hurting people I love.

Data Check: If you need more information, ask me questions before proceeding.

CHAPTER 15

Stealing and Other Terrible Ideas that can Teach Great Lessons

---

"My teenage son is taking things that don't belong to him and lying about it. He stayed with my parents for the weekend, and they had $20 on the kitchen table. When they went to get it, it was gone. They found it in his bookbag, and he said he didn't know who it belonged to. This is the second time; he also pocketed a wooden carving from a flea market. I made him return both stolen items and punished him. I know the dangers of stealing. How should I handle the situation?"

F irst, thank you for asking this question. You are not alone. Many parents face this painful moment, the one where your heart drops because your child has taken something that isn't theirs. When a child steals, it is not always about greed or defiance. It often reflects immature brain development, especially in the prefrontal cortex, which governs decision-making and impulse control. In simple terms, their "see, want, take" response fires faster than their conscience can intervene. The impulse system is fully operational while the control system is still under construction.

Parenting is leadership in its most personal form. In this case, you are the CEO watching a promising young employee violate company values and threatening the brand. It feels like betrayal, but it can become a powerful leadership moment. The true mark of a strong leader, whether in business or in family, is how they respond when things go wrong.

The first step is emotional intelligence before action. Take a deep breath. I know your mind may be racing with fear - visions of your child in a courtroom, handcuffed, or labeled forever by one bad decision. Those thoughts are understandable, especially for parents of children of color who know the stakes are higher. But fear-based leadership never works. When you lead from fear, your child will mirror your panic or retreat into shame. The goal is not to scare them straight but to guide them toward understanding and accountability.

Start with curiosity, not condemnation. Ask yourself, "What is driving this behavior?" Could your son be under stress? Is he angry, anxious, or feeling unseen? Great leaders look for system errors before assigning blame. In the same way, wise parents look beneath behavior to uncover need. You may find your child is not plotting deception but acting out confusion, loneliness, or peer pressure.

Once you have grounded yourself emotionally, shift to clear expectations and boundaries. Sit down together in a calm setting, not in the heat of the moment. Put away distractions and begin with empathy. Say, "I love you too much to ignore this. Stealing is not who you are, but it is something we must address." This statement does three things. It separates the behavior from your child's identity, sets the emotional tone, and establishes the coexistence of love and accountability.

Next, use "I" statements to express how the behavior affects you and others. "I felt hurt and disappointed when I learned you took Grandma's money. It made me question whether you understand how much trust matters in our family." This opens the conversation without judgment and models emotional regulation. You are teaching your child emotions can be expressed directly without aggression.

Ask reflective questions to invite problem-solving. "What were you feeling before you took the money? Did you think about who it belonged to? How did you feel afterward?" Encourage your son to describe his thoughts and emotions. Chances are, he has never slowed down long enough to process them. This is how emotional intelligence grows through reflection, not reprimand.

If you feel nervous about guiding the conversation, let AI help but use it intentionally. Write a prompt using your **ABCD Framework**, giving as much detail as possible in the Background section to receive personalized guidance. For example:

Prompt 88 - Parent: Empathy and Accountability Coaching

Actor: You are a child psychologist who specializes in teenage decision-making.

Background: My 15-year-old son has stolen twice, once from a family member and once from a store. I want to help him take responsibility, understand the emotional impact of his actions, and make amends.

Call to Action: Create a conversation script that teaches empathy, accountability, and self-regulation while maintaining a calm, respectful tone.

Data Check: If you need more information, ask me questions before proceeding.

This is how you incorporate AI wisely. Do not use shortcuts. Quick prompts produce shallow advice. When you think deeply about context, you invite AI to think deeply with you. The ABCD Framework is not busywork. It is your leadership structure for parenting with precision.

Once you have talked, move to accountability. Clear expectations and boundaries are the backbone of your family culture. State your family values out loud: "In this family, we practice honesty. We do not take what does not belong to us. We return what we find. We respect others' property, time, and trust." Repeat this message consistently. The goal is to make values visible.

Consider using empathy exercises. Ask him to imagine someone stole twenty dollars from his wallet. How would he feel? Let the silence stretch. Resist the urge to fill it. Learning happens in the pause. You might also use a role-reversal exercise. Borrow one of his favorite possessions without telling him. When he notices, discuss how it felt to lose something personal. Then explain gently, "That is what Grandma felt." End by returning it with transparency so he knows it was a teaching moment, not a trap.

When it is time for consequences, hold accountability with empathy. The punishment must fit the lesson, not just the offense. Require him to return the item in person and apologize. Reduce privileges that reflect responsibility, such as spending money or social outings, until he demonstrates consistent honesty. Make sure you say, "This is not about shame. It is about growth." Calm, consistent correction builds integrity more effectively than anger.

Now, model problem-solving. Show your son how to transform impulse into strategy. Talk about delayed gratification and goal-setting. "When you want something, we can make a plan to earn it." Teach him to pause before acting on impulse. Make waiting a challenge. "Let us see if we can wait five minutes before making a decision." Help him experience self-control as a strength, not a punishment.

AI can help you build this new skill. Use a savings app where he can track progress toward a goal. Have him use an AI budgeting assistant to plan how he will earn and save for what he wants. You are transforming temptation into training. Let him experience the satisfaction of earning instead of taking.

Bring your own transparency into the process. Share your experiences with temptation. "When I see something I want, I remind myself that earning it makes me proud. I wait until I can afford it." When your child sees you resist shortcuts, they learn what integrity looks like in practice.

If your son is old enough, involve him in drafting a family code of conduct. Make it collaborative. Ask AI for help wording it in kid-friendly language. Include statements like:

- We tell the truth even when it is uncomfortable.

- We respect what belongs to others.

- We take responsibility for fixing our mistakes.

Post this code somewhere visible. The purpose is not control; it is culture. Families should articulate their values clearly to build internal trust and external resilience.

Now, nurture a growth mindset. Remind your son this mistake does not define him. People are not their worst decisions. They are what they choose to do afterward. Say, "You made a bad choice, but you are still a good person. The next decision you make will shape who you become." Celebrate small wins. When he admits a mistake or returns something voluntarily, affirm it: "Admitting, your mistake took courage. I am proud of your honesty."

Invite your child to imagine his future self. "What kind of man do you want to be? How do you want people to describe you? Trustworthy? Responsible? Generous?" Encourage him to see character as a lifelong project. Growth mindset leadership teaches every setback is feedback, not failure.

If your son is a child of color, add context. Explain to him/her the world may not extend grace equally. Say, "It should not be this way, but it is. Your integrity is not optional. It is your protection." You are not scaring him; you are equipping him. You are preparing him to navigate a world, intentionally or not, often misinterprets youthful mistakes as threats.

Reinforce these values with service. Encourage generosity as the antidote to selfishness. Volunteer as a family. Let him experience how giving feels better than taking. Donate to causes together. Have him help younger children. Service teaches empathy in ways punishment cannot.

Reflect often on your own emotional tone. Anger may feel justified, but anger alone does not transform behavior. Steady, empathetic correction does. Children follow calm authority, not explosive reactions. Be the kind of leader your child will want to emulate: firm, fair, and forgiving.

End each conversation with hope. "I believe in you. I know you can make better choices." Leadership is about casting vision.

You are not only teaching honesty; you are teaching identity. You are showing your son he can recover from mistakes with grace and become a person of integrity and pride.

This is the long game of parenting leadership. You are raising a young man who must understand, real strength is in restraint and every action leaves an imprint on character. Stealing is a terrible idea, but handled wisely, it can become one of the greatest teaching moments of his life.

Your child is still learning what it means to lead himself. You are the leader teaching him how.

AI Prompts for Parents

Prompt 89 - Parent: Accountability Conversation Guide

Actor: You are a Parenting Coach

Background: My teen was caught stealing at school and lying about it.

Call to Action: I want to have a calm, honest talk that holds them accountable without shaming them. Also, provide me with 5 phrases I can use to start this conversation kindly but firmly.

Data Check: If you need more information, ask me questions before proceeding.

Prompt 90 - Parent: Impulse Control Strategies

Actor: You are a Child Development Specialist

Background: I want to teach my teen to resist instant gratification.

Call to Action: Strategies to help them delay impulses and make better choices. Suggest five (5) practical strategies to help my teen practice restraint, instead of giving in right away.

Data Check: If you need more information, ask me questions before proceeding.

Prompt 91 - Parent: Family Integrity Rules

Actor: You are a Family Communication Strategist

Background: I want to create clear family rules about honesty and respect for others' property.

Call to Action: Help me draft them in kid-friendly language. Give me five (5) simple, clear rules about honesty and stealing I can share with my child.

Data Check: If you need more information, ask me questions before proceeding.

AI Prompts for Young Leaders-In-Training

Prompt 92 - Student: Practicing Self-Control

Actor: You are an Emotional Intelligence Coach

Background: I want to be honest even when it's tempting to lie or steal.

Call to Action: Help me think before I act.

Data Check: If you need more information, ask me questions before proceeding.

Prompt 93 - Student: Practicing Financial Responsibility

Actor: You are a Goal-Setting Specialist

Background: I want things I can't afford right now.

Call to Action: Help me develop a plan to save or earn money to buy the things I want.

Data Check: If you need more information, ask me questions before proceeding.

Prompt 94 - Student: Moral Reasoning Exercise

Actor: You are a Self-Reflection Guide

Background: My friend's parents buy him anything he wants. He won't miss what I stole. I really wanted those trading cards.

Call to Action: Help me understand why stealing from someone who has a lot of everything, is wrong.

Data Check: If you need more information, ask me questions before proceeding.

CHAPTER 16

Name-Calling

"I recently learned my daughter has been calling her classmates and her younger siblings names. I have taken her phone and put her on restriction. Today, her gym teacher called and reported she called him a name under her breath. What should I do?"

By now, you should recognize the pattern, first, take a deep breath. As the Chief Emotional Officer (CEO) of your family, you have just learned one of your junior team members has been damaging the company brand with unprofessional communication. Before sending out an emotional disciplinary memo, or terminating the employee, pause and lead with strategy.

You are understandably frustrated. It can feel personal, as if your child has failed not only in manners but in character. But your real job is to help her learn words that have power, and leadership begins with emotional control. She needs to see you model calm authority while holding her accountable. Your message is simple: "This is serious, but it is fixable. I believe you can do better, and I am committed to helping you grow."

The first step is understanding the "why." Name-calling rarely comes from nowhere. It is often a reaction to frustration, embarrassment, insecurity, or peer pressure. Conduct your family version of a root-cause analysis before deciding on consequences. Ask yourself, "Is my child angry? Feeling excluded? Copying what she hears from others?" Behavior is a data point, not a diagnosis. You cannot fix what you do not understand.

Once you have gathered yourself emotionally, find a quiet time to talk. Turn off devices and sit together. Resist the impulse to lecture. Start with curiosity: "What was happening before you said that to your teacher?" or "What made you call your brother a name?" Listen without interruption. Even if her answer feels immature, hold space for it. When she says, "I don't know, I was just mad," respond with calm interest. "Okay, let's talk about why you were mad. Anger is normal, but hurting others with our words is not. Let's find better ways to express what you feel."

This moment is not just about discipline. It is your chance to teach emotional intelligence before action. You are showing your child how to handle strong emotions without letting them dictate behavior.

If you need support preparing for the conversation, use your ABCD Framework to create a reflective AI prompt. For example:

Prompt 95 - Parent: Coaching Respectful Communication

Actor: You are a child development specialist focused on emotional regulation.

Background: My daughter has been name-calling classmates and family members. I want to teach her to manage frustration respectfully and build emotional intelligence.

Call to Action: Create a conversation plan that helps me guide her toward empathy and accountability while keeping my tone calm.

Data Check: If you need more information, ask me questions before proceeding.

This approach helps you think deeply about context before seeking help. It ensures you use AI as a tool for reflection, not reaction.

Once you have set the emotional tone, establish clear expectations and boundaries. Every thriving organization has a code of conduct, and your family should too. Do not assume your daughter knows exactly where the line is. Spell it out clearly: "In our family, we do not call people names. Not teachers, not classmates, not siblings. We treat people with respect, even when we feel angry or annoyed." Avoid vague phrases like "be nice." Define respect in concrete, age-appropriate ways.

You can turn this lesson into a creative family project. Use AI to co-write a "family communication code." Invite her to contribute phrases or examples, and even design a poster together. When she helps shape the rules, she develops ownership. This transforms discipline into collaboration.

Now, move into modeling problem-solving. Children often resort to name-calling because they lack the vocabulary or tools to express strong emotions. Give her replacements. Practice phrases such as "I feel frustrated because..." or "I need space right now." Even humor can diffuse tension: "Why is everyone testing my patience today?" Encourage her to try these responses in different situations.

Practice is key. Just as an employee would not master public speaking after one seminar, your daughter will not master respectful communication after one talk. Role-play scenarios at dinner or in the car. Make it lighthearted. "Let's try this again, how could you say it differently?" Repetition builds skill and confidence.

As the leader of your home, remember that culture begins at the top. Reflect on your own language. Do you use sarcasm or snap under stress? Your daughter is watching and learning from you. If you occasionally slip, own it. Say, "I lost my patience earlier. What I said was not respectful. I am working on it too." Your humility will earn her respect far more than a lecture would.

Help her understand that name-calling is not strength, it is a sign of emotional immaturity. Explain, "You are smart. You have a powerful vocabulary. When you use insults, you are giving away your power. Real strength is choosing your words and showing self-control." Teach her that restraint is not weakness; it is leadership in action.

Now, guide her toward emotional regulation. Teach her to pause before reacting. "When you feel like saying something mean, take a breath. Ask yourself, 'What am I really feeling?' Then choose words that match the goal, not the emotion." You can use affirmations to reinforce this. Encourage her to repeat phrases like, "I can stay calm even when I am angry," or "My words show my strength." AI can help generate personalized affirmations or reminders.

When she slips, hold her accountable with empathy. Do not respond with sarcasm or name-calling in return. Say, "I am disappointed by what you said, but I respect you enough to speak calmly." This models exactly what you are teaching. Require her to make amends when appropriate. An apology to the teacher or sibling should include both reflection and a plan: "I was wrong. I let anger take over. Next time I will take a breath before I speak." Accountability delivered with kindness builds trust, not fear.

Positive reinforcement matters too. When she handles conflict well, celebrate it. "I noticed you were upset and chose not to call names, you showed real maturity." Praise effort, not perfection. You are shaping her internal feedback loop to value self-control over reaction.

Finally, nurture a growth mindset. Remind her that change takes practice. "You are learning how to use your words wisely. Every time you try again, you get stronger." Let her know that emotional regulation is a lifelong skill, not a one-time lesson. Share your own experiences with frustration. "I still have moments when I want to say something sharp, but I remind myself that calm communication gets better results."

Encourage her to use AI as her own reflection tool. She can practice difficult conversations with AI role-play prompts, such as, "Help me respond to a classmate who hurt my feelings without being mean." These rehearsals build muscle memory for respectful communication in real life.

If siblings are provoking her, address it. Hold every family member to the same communication standard. Remind them, "We respect each other, even when we disagree." Consistency keeps your family culture strong.

When she apologizes to the teacher, guide her to focus on integrity, not guilt. "Everyone makes mistakes, but what matters is how we repair them. Taking responsibility shows character." Then, when she follows through, acknowledge her growth. "You showed grace. Excellent leadership."

By staying calm, consistent, and compassionate, you are helping her transform name-calling into a masterclass on communication, empathy, and self-discipline. You are not simply correcting behavior; you are building character.

Remember, the goal is not to raise a child who never gets angry. The goal is to raise a young leader who knows how to express anger without cruelty, to stand firm without disrespect, and to use words as tools for connection, not destruction.

With you as the steady leader of your home and AI as your thinking tool, your family can turn even hurtful words into lessons on respect, self-control, and love in action.

AI Prompts For Parents

Prompt 96 - Parent: Respectful Communication

Actor: You are a Parenting Coach

Background: My teen uses profane insults with classmates and family.

Call to Action: I want to talk about it calmly but firmly, with help drafting my words. Help me use AI to generate five (5) clear, respectful phrases I can say to explain why name-calling isn't acceptable.

Data Check: If you need more information, ask me questions before proceeding.

Prompt 97 - Parent: Conflict Expression Skills

Actor: You are a Family Communications Strategist

Background: I want to teach my teen better ways to express frustration.

Call to Action: Suggest three (3) ways I can use AI role-plays or writing prompts to help my teen learn respectful language for conflict.

Data Check: If you need more information, ask me questions before proceeding.

Prompt 98 - Parent: Respectful Speech Accountability

Actor: You are an Emotional Intelligence Specialist

Background: I want to set clear family rules about respectful speech.

Call to Action: Help me write them in teen-friendly language using AI. Suggest five (5) simple, clear family communication rules for my family.

Data Check: If you need more information, ask me questions before proceeding.

AI Prompts For Young Leaders-in-Training

Prompt 99 - Student: Practicing Self-Control

Actor: You are an Emotional Intelligence Coach

Background: I say mean things and call names when I'm angry.

Call to Action: I want to learn to stop myself. How can I avoid saying unkind things.

Data Check: If you need more information, ask me questions before proceeding.

Prompt 100 - Student: Respectful Emotional Expression

Actor: You are a Communication Specialist

Background: I want to say I'm mad without cursing or name-calling.

Call to Action: Help me express myself and communicate how I am feeling without being disrespectful. Especially when I am angry.

Data Check: If you need more information, ask me questions before proceeding.

Prompt 101 - Student: Emotional Regulation Strategies

Actor: You are a Self-Reflection Guide

Background: When I am angry, I can't think clearly, and I just say the first thing that comes to mind.

Call to Action: How can I avoid losing control? What are some strategies to help me process my emotions in an emotionally mature manner?

Data Check: If you need more information, ask me questions before proceeding.

CHAPTER 17

Developing Resilience

"My child gives up too easily, in discussions, in sports, even in her coursework. I want to help her develop greater capacity for resilience. How do I teach her not to quit so easily?"

First, let us acknowledge the importance of your observation. You have spotted something many parents miss until their child is much older: a fixed mindset. When a child believes success comes only from natural talent instead of practice and persistence, they begin to avoid challenges that might reveal weakness. Catching this early gives your child the greatest chance to grow into a strong, capable, and confident leader of her own life.

Resilience begins with an emotional tone. Quitting is often not about laziness; it is about fear - fear of failure, embarrassment, or disappointing others. Before you correct her behavior, create a safe space for her to name those fears. Say, "I understand. It can feel hard to keep going when it seems like you are not winning or don't know what to do. I have felt that too." By modeling vulnerability, you teach her courage is not the absence of fear but the decision to move forward in spite of it.

Next, explore what drives her behavior. Does she lack confidence, patience, or problem-solving tools? Does she quit because relief feels easier than perseverance? Understanding the reason helps you guide, not just discipline. Sometimes professional support from a counselor or psychologist can uncover deeper issues such as anxiety, perfectionism, or fear of judgment. When you know the root cause, your coaching becomes targeted instead of reactionary.

If her struggle is confidence, start with small wins. Give her achievable tasks that help her experience success through completion. If frustration is the barrier, teach her to take short breaks when overwhelmed and return to the task refreshed. And if fear of failure drives her behavior, normalize mistakes as information, not identity. Help her reframe failure as a teacher rather than a verdict.

Bring storytelling into your teaching. Children learn resilience by seeing it lived. Share stories of leaders who failed before they flourished. Tell her about Michael Jordan being cut from his high school basketball team, J.K. Rowling's twelve rejections before publishing Harry Potter, or Beyoncé losing on *Star Search* before becoming one of the world's most disciplined performers. Explain every great success has been built on thousands of imperfect attempts. Then share your own stories of persistence. When she sees your humanity, she realizes setbacks are not the opposite of success, they are the pathway to it.

You can use AI as your leadership tool in this process. Create prompts to help you plan intentional conversations. For example, ask, "Give me ten ways to explain the value of perseverance to a ten-year-old," or "Help me design a family mantra about finishing what we start." When you write the prompt in your **ABCD Framework,** include context: Your child's age, the situation, and the emotional tone you want to model. AI can help you transform "try harder" into actionable strategies to teach persistence step by step.

Resilience grows through problem-solving, not pep talks. Instead of saying, "Don't quit," Help her explore *why* she wants to keep going. Say, "I noticed you stopped. What was happening at that moment?" She may reveal embarrassment, boredom, or confusion. These moments are gold. They give you insight into her thought process and help her see that discomfort is part of progress.

Once you understand her trigger, teach her how to manage it. If she quits because she feels stuck, break the task into smaller steps. If she fears judgment, help her develop self-talk such as, "It's okay to make mistakes while learning." Encourage her to use AI to create affirmations or motivational scripts for herself. The goal is to shift from emotional reaction to strategic thinking.

As a parent leader, your words carry the power to shape your family's expectations. Make your family values about effort explicit: "In this family, we finish what we start." Explain that following through is an act of integrity, not perfection. Clarify that persistence is how we honor commitments to ourselves and others. Then role-play situations where quitting might be tempting. Walk through better options: taking a break, asking for help, or changing approach. Practice these in real time so when challenges appear, she has tools ready.

AI can also serve as your rehearsal coach. In private moments, use your prompts to practice how you will respond when your daughter wants to give up. For example:

Prompt 102 - Parent: Growth Mindset Coaching

Actor: You are a sports psychologist for middle school athletes.

Background: My eleven-year-old daughter gives up easily when she makes mistakes during soccer practice. I want to respond with calm encouragement instead of frustration.

Call to Action: Create a short script I can use, in the moment, to help her re-engage without pressure.

Data Check: If you need more information, ask me questions before proceeding.

Practicing with AI keeps your leadership voice composed and consistent. It prepares you to respond with empathy rather than exasperation when the moment arrives.

Once you have established a supportive tone, help your child strengthen her persistence muscles through structure. Set clear, realistic goals together and track progress visually. Create a "Resilience Board" or digital tracker where she can list challenges and record when she overcomes them. Celebrate effort, not only outcomes. Say, "You didn't quit when it got hard; that's what matters most."

Incorporate growth mindset language into daily life. Replace "You are so smart" with "I love how you kept working on the problem." Encourage effort by saying, "You practiced until it became easier." Praise persistence, curiosity, and creativity. These words build identity around effort and adaptability instead of perfection.

Invite your daughter to use AI to explore resilience stories that inspire her. Ask AI for book lists or short biographies about people who overcame setbacks. Turn resilience into a shared learning journey. During dinner, talk about small moments from your day when you almost gave up but pushed through. Let her see resilience modeled as part of everyday life.

And when she does give up, resist shaming her. Meet the moment with empathetic accountability. Say, "It looks like you stopped. What happened?" Hold her accountable to reflect and plan. "What could you try differently next time?" This helps her move from avoidance to ownership. Over time, these conversations teach her persistence is a choice she can make again and again.

Encourage her to embrace what I call "micro-comebacks", moments where she restarts even after stopping. When she walks away from a task, gently invite her to return later. The goal is not perfection but recovery. Quitting becomes a pause, not an endpoint.

And when she finishes something that once felt impossible, celebrate the process. "You didn't quit when it got tough, and now look at what you have accomplished." Recognition reinforces the emotional reward of perseverance.

Ultimately, resilience is not just about grit it is about grace under pressure. It is learning to pause, recalibrate, and try again. It is knowing failure does not define you but refines you. You are helping your child build the emotional and cognitive muscles needed to transform frustration into focus.

With emotional intelligence to guide your tone, clear expectations to frame your standards, problem-solving to sustain momentum, empathy to keep accountability humane, and a growth mindset to power the long game, you are not just teaching resilience, you are modeling it.

Because resilience is not inherited. It is practiced. And with you as her example, and AI as a thoughtful planning tool, your daughter will not just learn to try again she will learn to rise again.

AI Prompts For Parents

Prompt 103 - Parent: Growth Mindset Training

Actor: You are a Parenting Coach

Background: My child gives up easily on tasks or social challenges. I am frustrated my child lacks follow-through. My child is [enter here] years old and in the [enter here] grade.

Call to Action: How can I coach my child calmly and avoid shaming them? How can I help them learn to keep going, even when it is hard?

Data Check: If you need more information, ask me questions before proceeding.

Prompt 104 - Parent: Family Perseverance

Actor: You are a Family Strategy Assistant

Background: I want to set clear family expectations about perseverance. My child is [enter here] years old and in the [enter here] grade.

Call to Action: Create a plan to help me guide my family in the development of family agreements about our family's commitment to finishing what we start. And, how to ask for help.

Data Check: If you need more information, ask me questions before proceeding.

Prompt 105 - Parent: Growth Mindset Coaching

Actor: You are an Emotional Intelligence Specialist

Background: I want to help my child learn how to motivate themselves to keep working when they struggle with their homework. My child is [enter here] years old and in the [enter here] grade.

Call to Action: Provide guidance on how to coach my child through completing more challenging homework.

Data Check: If you need more information, ask me questions before proceeding.

AI Prompts For Young Leaders-In-Training

Prompt 106 - Student: Growth Mindset

Actor: You are a Resilience Coach

Background: I want to quit when things get hard. I am in the [enter here] grade and I am [enter here] years old.

Call to Action: I want help learning to keep going.

Data Check: If you need more information, ask me questions before proceeding.

Prompt 107 - Student: Homework Help

Actor: You are a Communication Assistant

Background: Sometimes when I do my homework, I don't know the answer and my Mom gets angry with me. I am in the [enter here] grade and I am [enter here] years old.

Call to Action: How do I ask for help from my Mom.

Data Check: If you need more information, ask me questions before proceeding.

Prompt 108 - Student: Study Success Plan

Actor: You are a Self-Reflection Guide

Background: In class, I understand the teacher's instructions. But, when I do my homework at home, I can't remember the instructions. I am in the [enter here] grade and I am [enter here] years old.

Call to Action: Help me understand why I can't complete my homework. How can I get better at completing my homework?

Data Check: If you need more information, ask me questions before proceeding.

CHAPTER 18

New Soil, Same Roots

"My kids really don't want to move. We are downsizing to control expenses, which means a new school, new friends, and a smaller space. I don't have a choice, but my children are upset about the move. How do I help them make the transition?"

First, let us pause and acknowledge your courage. You are making a decision that requires strength, foresight, and leadership. Choosing financial stability over comfort is not an easy choice, yet it is one of the most responsible actions a parent can take. When a family's expenses outgrow its income, stress and tension begin to take root. By making this move now, you are protecting your family's future and modeling wise leadership under pressure.

Your children may not see it that way. They do not see bank statements or balance sheets. They see the bedroom they are leaving behind, the friends they will miss, and the routines that make them feel safe. For them, this transition represents loss. Your role as the leader of your home is to help them see change is not an ending but a new beginning. Growth happens when we are willing to be replanted in new soil.

Begin by setting the emotional tone. Your calm presence will shape how your children experience the move. If you present it with anxiety or anger, they will mirror those emotions. If you approach it as an opportunity to grow stronger as a family, they will eventually follow your lead. Speak honestly but confidently. You might say, "I know this is hard for all of us. I feel sad about leaving too, but I am proud of us for making a smart choice and keeping our family secure." Your composure teaches your children courage is not about being fearless. It is about acting with purpose even when it is uncomfortable.

Before packing a single box, plan a family meeting. Use your **ABCD Framework** to create an AI prompt that helps you structure the conversation.

Prompt 109 - Parent: Family Transition Conversation

Actor: You are a family leadership coach guiding parents through major transitions.

Background: We are moving to a smaller home to reduce expenses. My children are nervous and upset.

Call to Action: Help me create a calm and hopeful script for explaining the move and addressing their concerns with empathy.

Data Check: If you need more information, ask me questions before proceeding.

This kind of structured prompting helps you prepare your message in advance so emotions do not cloud your delivery. You are using AI not to replace your leadership but to strengthen it.

Next, establish clear expectations and boundaries. Change often feels overwhelming because it removes predictability. Giving your children structure helps them feel anchored. Make it clear the move is a family project, not a punishment. Share with them, "Everyone in our family has a role to play in this move. We are working together to make our new home comfortable and peaceful." Assign responsibilities that match their age. Younger children can help choose toys to donate or place stickers on boxes. Older children can plan menus, organize their own rooms, or research local parks and clubs. When every person has a clear role, the transition becomes shared teamwork rather than forced compliance.

Expect resistance and meet it with empathy. When a child says, "I hate this idea" or "I don't want to move," do not dismiss their feelings. Respond with curiosity. "Tell me what worries you most about moving. Let's talk about it together." They may express fears about losing friends or not fitting in at a new school. Listening deeply validates their emotions and creates a sense of safety. Remember, emotional intelligence before action always builds trust.

Use AI to help design meaningful rituals honoring both the old and the new. Ask, "What are creative ways to help children say goodbye to a home they love?" The response might include taking photos in favorite rooms, writing thank-you notes to the house, or hosting a family dinner where everyone shares their favorite memory. These acts of closure transform sadness into gratitude. They show your children that endings can be handled with grace.

Once the emotions have been acknowledged, shift into problem-solving. Showing your children adapting to change is a learnable skill. When they express uncertainty, say, "We can figure this out together." Model adaptability through small actions. When you cannot find something after packing, laugh and say, "We will discover it soon enough." When you are unsure of the route to your new grocery store, invite them to help navigate. "Let's see who can find it first." Turning the unknown into an adventure builds flexibility and teamwork.

Teach your children resilience grows in unfamiliar soil. Change stretches our capacity to think creatively, build new relationships, and discover hidden strengths. Remind them every person who achieved something remarkable did so by embracing the unfamiliar. Moving to a new home is not losing what you had; it is expanding what is possible.

This is where empathetic accountability matters most. Set boundaries around how emotions are expressed. Say, "You are allowed to be sad or angry. You are not allowed to be unkind or refuse to participate." Boundaries teach that feelings are valid but must be expressed with respect. Maintaining structure during upheaval gives children emotional security.

You will likely encounter moments of protest. Tears, slammed doors, or silent resistance are part of the process. Respond with calm and consistency. "I understand you are upset, and it is okay. Let's keep working together." Your steadiness teaches them leadership means showing up even when it is hard. Over time, your composure becomes the anchor that helps them adjust.

Communication is your greatest ally. Schedule regular family check-ins where everyone can express how they are feeling about the move. Ask, "What has been challenging this week? What has been better than you expected?" Listen carefully and thank them for sharing. These meetings remind your children their voices matter and that they are part of a team navigating change together.

Help your children reframe the story of the move. Instead of saying, "We have to move," say, "We get to start fresh." Emphasize the benefits of simplicity, teamwork, and new experiences. Remind them home is not defined by square footage but by connection and love. "Our family's strength moves with us. Wherever we go, we bring our kindness, our laughter, and our traditions."

Use AI as a creative instrument to discover ways to rebuild community. Ask for ideas to find local libraries, clubs, or volunteer activities that match your children's interests. Invite them to choose one activity they want to try. This helps them invest emotionally in their new environment and feel a sense of agency over their future.

Recreate familiar family rituals in your new space. Keep your movie nights, Sunday breakfasts, or evening walks. These small continuities reassure your children even though the walls have changed, the heart of your family remains the same.

Most importantly, teach them to embrace change instead of fighting it. Growth requires movement. Change is not the enemy; resistance is. The more your children learn to approach new experiences with curiosity and confidence, the stronger and more adaptable they become. Explain how change is like repotting a plant. It can feel uncomfortable at first, but the roots grow deeper and stronger when given room to expand.

Reflect on your own mindset too. Speak openly about how you are adjusting. "I miss some things about our old home, but I am learning to enjoy our new neighborhood. I am proud of us for being brave." This kind of transparency models emotional regulation and a growth mindset. Your children will learn transformation takes time but always leads to renewal.

Through emotional intelligence, clear expectations, collaborative problem-solving, empathetic accountability, and a growth mindset, you are not just managing a move. You are teaching your family how to lead through life's inevitable transitions. You are showing them change is not something to survive but something to embrace, because it carries the promise of new strength and new beginnings.

Home is not the building you live in. It is the love, resilience, and leadership you bring into every space you occupy. Your family is not losing its roots. You are simply planting them in new soil where they can grow stronger and reach higher together.

AI Prompts For Parents

Prompt 110 - Parents: Managing Change

Actor: You are a Parenting Coach

Background: My children are resistant to moving to a smaller house. My children are ages [enter here].

Call to Action: How do I explain the reasons for the move without scaring my children?

Data Check: If you need more information, ask me questions before proceeding.

Prompt 111 - Parent: Shared Responsibility

Actor: You are a Family Planning Assistant

Background: I want to involve my kids in the moving process.

Call to Action: What are some ways I can give them responsibility without overwhelming them?

Data Check: If you need more information, ask me questions before proceeding.

Prompt 112 - Parent: Managing Change

Actor: You are an Emotional Intelligence Specialist

Background: I want to talk with my kids about their fear of not making new friends.

Call to Action: Provide me with a script for the conversation.

Data Check: If you need more information, ask me questions before proceeding.

AI Prompts for Young Leaders-in-Training

Prompt 113 - Student: Relationship Management

Actor: You are a Resilience Coach

Background: I'm scared about moving and leaving my old friends. I am in the [enter here] grade and I am [enter here] years old.

Call to Action: How can I keep my old friendships?

Data Check: If you need more information, ask me questions before proceeding.

Prompt 114 - Student: Relationship Management

Actor: You are a Communication Assistant

Background: I'm worried about meeting new people at my new school. I am in the [enter here] grade and I am [enter here] years old.

Call to Action: Help me figure out what to say and how to make new friends.

Data Check: If you need more information, ask me questions before proceeding.

Prompt 115 - Student: Coping with Change

Actor: You are a Self-Reflection Coach

Background: I feel sad and angry about moving. I am in the [enter here] grade and I am [enter here] years old.

Call to Action: How can I get through my move that I don't want?

Data Check: If you need more information, ask me questions before proceeding.

CHAPTER 19

Redefining the Family Blueprint

""I'm going through a difficult divorce and worried about how it's affecting my kids. How do I help them handle this massive change?"

F irst, take a deep breath. You are walking through one of the hardest transitions any family can face. Divorce can shake the foundation of what "family" means. Yet, it also creates an opportunity to redefine meaning with intention, care, and clarity. You are not losing a family. You are rebuilding one with purpose.

As the leader of your household, much of the emotional and practical burden will fall on you. There will be days when the weight feels unbearable. Leadership during this time requires presence. Even when you are tired, your consistency gives your children a sense of safety.

Begin by setting the emotional tone. Avoid pretending everything is fine or framing this as an exciting adventure. Children see through pretense. Instead, model calm honesty. You might say, "This is going to be challenging in the beginning. We are all going to have big feelings, and that's okay. We can talk anytime about anything. We are going to get through this together." Your children need to hear their emotions are valid and you can handle them.

Emotional tone also means showing your humanity. Many parents try to appear strong by hiding their pain, believing tears will scare their children. Yet, showing appropriate emotion teaches resilience. When children see both your sadness and your strength, they learn it is possible to hurt and still move forward. After my own divorce, my children's therapist told me my composure was making them worry that I did not care. Once I allowed them to see my sadness, they relaxed. They learned courage does not mean holding back tears. It means feeling deeply and still standing tall.

Use AI to help you find words when emotions are raw. You might ask, "How can I explain divorce to a six-year-old?" or "Give me five ways to respond when my child asks a question I do not know how to answer." Prompts like these prepare you to approach difficult conversations with compassion and confidence instead of fear or frustration.

Divorce often makes life feel unstable for children. They may test limits, use guilt to negotiate, or lash out at the parent they live with most. This behavior is not disrespect; it is disorientation. The best way to restore balance is through consistency. Keep your family's code of conduct in place. Say, "Even though things are changing, our values are not. We still respect each other. We still tell the truth. We still take responsibility for our actions." Boundaries build safety. Even in chaos, they communicate, "You can trust me to lead."

Use AI to co-create simple family agreements and define expectations clearly and kindly. Ask, "Help me create a family communication agreement for children adjusting to divorce." Provide context in your prompt so AI can generate language tailored to your children's ages and personalities. Post this agreement where everyone can see it. It serves as both a reminder and as reassurance.

Divorce brings practical and emotional challenges. Two homes, two sets of rules, and a schedule that feels unfair can overwhelm even the most adaptable child. When they complain, "I don't want to go to Dad's house," or "I hate going back and forth," resist the urge to end the discussion quickly. Say, "Tell me what feels hardest. Let's think of what you can try next time." This approach transforms frustration into problem-solving. It invites emotional honesty without judgment.

At times, guilt will tempt you to become too lenient, while exhaustion may push you toward harshness. Both extremes are counterproductive. What your children need most is empathetic accountability. Hold them to clear expectations with love. "I know this is frustrating, but our routines are still important." When they tantrum or withdraw, name the emotion and provide stability. "You are angry because this feels unfair, but you still must pack your bag for tomorrow." Calm repetition of clear messages builds security over time.

You can also prepare for tough moments in advance. Create AI prompts that help you practice responses to tears, anger, or silence. For example: "Act as a family therapist and write three supportive phrases I can use when my child cries about visiting the other parent." Practicing calm communication through AI helps you stay centered when emotions run high.

Divorce is not just an ending; it is a redesign. It is a chance to build a new family structure based on truth and mutual respect. Nurturing a growth mindset by showing your children change, though painful, can lead to strength and wisdom is important and necessary. Tell them, "This is hard right now, but it will not always be this way. We are learning how to handle hard things." Encourage them to imagine what your new normal can look like more one-on-one time, new traditions, and fresh routines they can help design.

Let your children see you learning, too. When you make mistakes, own them. "I am sorry for snapping earlier. I am tired and upset, but that is not fair to you." Apologizing does not weaken your authority; it strengthens it. It models humility, repair, and emotional intelligence. It teaches love is not about perfection but about effort and responsibility.

Use AI intentionally to fill emotional and logistical gaps. Ask for suggestions on maintaining consistency across two homes or ideas for shared family calendars that reduce confusion. Create packing checklists so your children remember important items when switching homes. This kind of structure prevents unnecessary tension and builds confidence in your ability to manage change together.

Keep routines intact. Bedtimes still matter. Chores still matter. Family dinners still matter. These rituals are not luxuries, they are anchors. When life feels uncertain, structure communicates safety. Tell your children, "Even though many things are changing, some things will always stay the same." Routine becomes a language of love and it says, you can count on me.

Now, help your children redefine what "family" means. Say, "Our family looks different now, but it is still a family. We are going to figure out what it means together." Ask for their input. "What new traditions should we create? How can we stay connected when we are apart?" Listen to every idea, even the unrealistic ones. Their voice in this process builds agency and emotional security.

AI can help you facilitate these family conversations. Ask for agendas to keep meetings supportive and balanced. Request scripts that make discussions age-appropriate for everyone from the teenager to the first-grader. Use AI to design a "Family Agreement" about respect, communication, and shared responsibilities. Post it on the refrigerator as a daily reminder of your collective commitment.

Continue holding your children accountable for how they treat others and themselves. Divorce is not an excuse to let your family's standards slip. When disrespect surfaces, respond calmly: "I understand this is difficult, but we still speak with kindness." This teaches them integrity does not depend on circumstances; it is a choice made every day.

Encourage a growth mindset in small, practical ways. When your child says, "This is too hard," reframe it gently: "It is hard right now, but you are learning how to handle hard things." Celebrate effort rather than perfection. Praise the courage to try again, the honesty to express emotion, and the resilience to adapt. These moments of affirmation shape character far more than lectures ever could.

Use AI to gather ideas for books and movies that show families overcoming change with grace. Discuss these stories together. Ask, "What did the character learn about family?" or "How did they stay strong even when things were hard?" These reflections help children see themselves as capable of navigating complex emotions.

Most importantly, remember your children do not need a perfect home. They need a safe one. They do not need a flawless parent. They need a present one. Through emotional intelligence, clear expectations, collaborative problem-solving, empathy, and a growth mindset, you are proving leadership at home is not about maintaining the old blueprint. It is about redesigning it with courage, compassion, and care.

Family is not about being under one roof. It is about showing up, day after day, with honesty, patience, and love. Divorce may change the shape of your family, but it cannot erase its foundation. What you are building now is something new, grounded in truth and nurtured by hope.

AI Prompts For Parents

Prompt 116 - Parent: Divorce Conversation Guide

Actor: You are a Parenting Coach

Background: I'm trying to talk with my kids about our divorce without scaring them. The divorce process has been [enter here] and they are struggling with [enter here]. Their ages are [enter here] and they are in grades [enter hee].

Call to Action: Suggest strategies for talking to my children about divorce.

Data Check: If you need more information, ask me questions before proceeding.

Prompt 117 - Parent: Family Meeting Plan

Actor: You are a Family Communications Coach

Background: I want to hold family meetings about our new schedule and expectations. The divorce process has been [enter here] and they are struggling with [enter here].

Call to Action: Help me plan age-appropriate family meetings.

Data Check: If you need more information, ask me questions before proceeding.

Prompt 118 - Parent: Family Expectations Framework

Actor: You are an Emotional Intelligence Specialist

Background: I want to set clear expectations about behavior during this transition. The divorce process has been [enter here] and children are struggling with [enter here].

Call to Action: Help me identify a set of expectations in kid-friendly language. Provide me with a list of examples to choose from.

Data Check: If you need more information, ask me questions before proceeding.

AI Prompts for Young Leaders-in-Training

Prompt 119 - Student: Talking Through Change

Actor: You are a Resilience Coach

Background: I feel [enter here] about my parents' divorce. I am [enter here] years old and in the [enter here] grade.

Call to Action: Suggest ways to talk to my parent(s) about my feelings.

Data Check: If you need more information, ask me questions before proceeding.

Prompt 120 - Student: Finding Balance Between Homes

Actor: You are an expert at helping family through transitions.

Background: I have to switch houses and I don't like it. It is hard to adjust to different rules and different routines every week. I am [enter here] years old and in the [enter here] grade.

Call to Action: Help finding ways to cope with the constant change.

Data Check: If you need more information, ask me questions before proceeding.

Prompt 121 - Student: Facing Fear with Courage

Actor: You are a Child Therapist

Background: I am afraid of [enter here]. I am [enter here] years old and in the [enter here] grade.

Call to Action: Suggest ways to cope with my fear about [enter here].

Data Check: If you need more information, ask me questions before proceeding.

CHAPTER 20

The Silent Shift

"My 14-year-old son has suddenly changed. He used to be outgoing and happy. Over the last couple of months he's become very secretive. He hides his phone when I sit near him and his grades are slipping. Should I be concerned!"

I t is completely appropriate to be concerned about a gradual change in your child's behavior. When a once-engaged child becomes distant, withdrawn, or secretive, it signals the need for closer attention and care. Start with calm curiosity instead of alarm. Emotional intelligence before action allows you to notice what is really happening beneath the surface rather than reacting out of fear.

Adolescence can be a storm of hormones, peer dynamics, and social pressures. Teens often retreat when they feel uncertain or overwhelmed. Pay attention to other shifts. Has he changed friend groups, started avoiding activities he once loved, or withdrawn from family life? Has his sleep, appetite, or hygiene changed? Does he seem fatigued or complain about headaches and stomach pain? Have his teachers noticed changes in class participation or mood? These clues help you discern whether you are dealing with typical teen turbulence or a deeper concern.

If there is no clear medical reason for the change, you will need to engage more intentionally. Begin with an open, calm conversation. Write an AI prompt to generate supportive, nonjudgmental language such as, "I've noticed you seem quieter lately and your grades have dropped. How are you feeling?" Another example could be, "I've noticed something feels different. I am not here to scold. I am here to listen." Your tone is everything. When your child senses empathy rather than accusation, he is more likely to open up.

As the leader of your household, your first duty is to ensure safety. Schedule an appointment with your child's pediatrician and a licensed therapist to rule out medical or emotional causes. Stay engaged with teachers, counselors, and coaches. Build what I call a "community of care," a network of trusted adults who can support your son's well-being.

Set clear expectations and boundaries around technology. You are the Chief Executive Officer of your family's enterprise, and which means safeguarding the digital environment in which your child spends time. Create AI prompts to help you identify modern parental control tools, phone monitoring apps, and strategies for teaching online safety. Ask AI to help you explain why you are setting these limits in a calm and factual way. You might say, "I love you and respect your need for privacy, but part of my job is to keep you safe online. I will check your devices and talk with you about what I find. You deserve protection and guidance, not secrecy."

If you find alarming content such as bullying, drug use, or contact with strangers, act immediately. This is not a time for silence. Contact school officials, counselors, or even law enforcement if necessary. Tell your child directly, "If someone is asking you to do things and you are uncomfortable, I will step in to protect you, even if that means getting help from others." Accountability rooted in love teaches leadership sometimes requires courage in uncomfortable moments.

When you speak with your child, manage your emotions carefully. Setting the emotional tone means using your calm voice even when you are anxious. Avoid barging in with anger or interrogation. Say, "I have noticed some changes, and I need your help to understand what is going on." This communicates partnership rather than punishment.

Be prepared for resistance. Teenagers value autonomy, and they may react defensively when boundaries tighten. Stay firm but kind. Say, "I know you want privacy. I want it for you too, but right now, some of your choices are causing concern. We will work together to rebuild trust." Enforce boundaries consistently. If screen privileges are removed, explain clearly why and what must happen for them to be restored. Consistency communicates love more powerfully than lectures ever could.

Use your **ABCD Framework** as a family communication guide. Begin with the actor: "You are my son, and I love you." Then move to the background: "I have noticed changes in your grades, your mood, and how you spend time on your phone." Provide the call to action: "Help me understand what is happening so we can figure this out together." End with the data check: "If I am missing something, tell me what you want me to know." This structure keeps the conversation constructive instead of confrontational.

If your child struggles to articulate his feelings, let AI assist in designing reflective exercises. Ask for conversation starters like, "Give me five ways to help a teenager talk about emotions when they shut down." Encourage your child to use AI himself to practice expressing thoughts before sharing them aloud. You are using technology not to invade, but to support communication and understanding.

Problem-solving may also mean adjusting routines. If your son needs internet access for school, have him complete homework in a shared area like the kitchen table. Transparency encourages accountability. Make it clear supervision is not punishment; it is partnership. Say, "I trust you want to make good choices. Let's work together to rebuild that trust."

As you move through this process, stay anchored in empathetic accountability. When your son resists, do not withdraw. Show affection while holding your position. His prefrontal cortex is still developing, which means impulse control and risk assessment are still under construction. He needs a steady adult who can lead with both firmness and grace.

Encourage healthy outlets that promote belonging and self-esteem. Sports, music, art, volunteering, or faith-based groups can provide purpose and community. Involvement in positive activities replaces isolation with engagement. Celebrate effort, not just outcomes. "I am proud of how you showed up for practice even though you did not feel like it." Every word of encouragement builds resilience.

Nurture a growth mindset by reframing setbacks as opportunities to learn. When grades slip, avoid shame-based comments. Instead say, "This grade shows where you are right now, not who you are. Let's figure out what to do next." Focus on the process of improvement rather than punishment. Praise perseverance, curiosity, and effort.

Most importantly, continue to show up. Parenting through silence and secrecy requires patience and emotional stamina. Remind yourself that leadership at home is not about control. It is about presence. It is about staying close even when your child pulls away. Say often, "You can always come to me, no matter what."

AI can help you stay informed and equipped, but your warmth, intuition, and consistency are what truly guide your family through this season. When you blend empathy with clear boundaries, accountability with compassion, and firmness with flexibility, you are modeling the very skills your child will one day need to lead himself.

The silent shift you see in your son is not just a sign of trouble. It is a moment of transformation. With your guidance, steady emotional tone, problem-solving mindset, and leadership grounded in love, this period can become a bridge to deeper trust and maturity. Every family faces moments of silence. What defines strong families is how they choose to listen, respond, and grow through them.

AI Prompts For Parents

Prompt 122 - Parent: Digital safety Check

Actor: You are a Cyber Safety Specialist

Background: My teen is being very secretive and his behavior has changed. I am concerned his technology may have enabled "bad actors" to engage with him. My son is [enter here] years old and has begun to [enter here].

Call to Action: How do I investigate his phone and computer for unsafe digital interactions? Create a detailed plan of action with step by step instructions to examine his technology.

Data Check: If you need more information, ask me questions before proceeding.

Prompt 123 - Parent: Respect & Safety

Actor: You are a Family Therapist

Background: My teen is being very secretive and his behavior has changed. I need to explain why I'm checking my child's phone.

Call to Action: Provide me with a script to use with my child that balances respect with safety.

Data Check: If you need more information, ask me questions before proceeding.

Prompt 124 - Parent: Investigating Secretive Behavior

Actor: You are a Child Psychologist

Background: My teen is being very secretive and his behavior has changed. Over the last couple of months he's become very secretive. He hides his phone when I sit near him and his grades are slipping.

Call to Action: How should I address his recent behavior? Provide me with different strategies.

Data Check: If you need more information, ask me questions before proceeding.

AI Prompts for Young Leaders-in-Training

Prompt 125 - Student: Asking for Space

Actor: You are an Emotional Intelligence Coach

Background: I've been feeling stressed and want some privacy. I am in the [enter here] grade and I am [enter here] years old.

Call to Action: Help me figure out when it is okay to handle my feelings on my own and when I should talk to my parents about what is going on.

Data Check: If you need more information, ask me questions before proceeding.

Prompt 126 - Student: Requesting Privacy

Actor: You are a Communication Assistant

Background: My [Mom/Dad/Guardian]wants to check my phone and I feel like that's an invasion of my privacy. I am in the [enter here] grade and I am [enter here] years old.

Call to Action: Help me express my feelings about privacy and trust in a respectful way so [he/she/they] understand my point of view without it turning into an argument.

Data Check: If you need more information, ask me questions before proceeding.

Prompt 127 - Student: Independence vs. Safety

Actor: You are a Self-Reflection Guide

Background: I really feel like I have a right to privacy. I am in the [enter here] grade and I am [enter here] years old.

Call to Action: Help me explore why my parents might feel worried about my privacy, what they are trying to protect me from, and how I can balance my need for independence with their concern for my safety.

Data Check: If you need more information, ask me questions before proceeding.

CHAPTER 21

Modeling Generosity in a Selfie World

"My teen children do not recognize my birthday and holidays. Even if I give them money to buy me a gift, they will buy something cheap and keep the excess. On my last birthday, they just sent a birthday text. But, they expect me to go all out for them. They don't even get me a decent Christmas present with my money. What have I done wrong?"

Take a breath. You have not failed beyond repair. Most parents who find themselves here are not neglectful; they are generous to a fault. You have spent years making birthdays, holidays, and milestones unforgettable for your children. You have anticipated their needs before they even asked, pouring your time, creativity, and resources into showing them love. Somewhere along the way, they learned to accept your efforts as a given rather than as acts of care. They have gotten used to you carrying the emotional labor of celebration while they simply receive. This is not cruelty or indifference. It

189

is a learned pattern. The good news is your kids are capable of learning new patterns.

Begin by setting the emotional tone. This is not about shaming your children for selfishness but about modeling the kind of empathy and generosity you want them to develop. Sit down with them and speak plainly: "It hurts my feelings when my birthday is treated like it doesn't matter. I want us to be a family willing to celebrate each other." Your tone should be steady and kind, not guilt-driven. This conversation is an act of leadership. You are modeling emotional honesty by expressing what you need without accusation. When children see adults can articulate hurt feelings calmly and respectfully, they begin to understand that vulnerability is a form of strength, not weakness.

Next, establish clear expectations and boundaries. Many children treat a parent's disappointment as optional because the consequences are minimal. Clarify your family's standards for reciprocity. Say, "In this family, we celebrate one another. I expect you to show thought and effort for birthdays and holidays. It does not have to be expensive, but it must come from the heart." Be specific about what effort looks like. Describe actions such as a handwritten note, cooking a favorite meal, planning a small outing, or saving money to buy something meaningful. The goal is not to control their choices but to teach love requires intention.

Problem-solving means guiding them on how to give, not just telling them to do it. Many teens are not unkind, just inexperienced. They do not know how to plan ahead, manage a budget, or pick a meaningful gift. Help them brainstorm ideas. Invite them to use AI to come up with creative and affordable gifts or gestures. Let them see technology can support generosity instead of replacing it. They can use prompts like,

"What are thoughtful birthday gifts under twenty dollars for someone who loves reading and gardening?" or "Help me write a heartfelt note for my mom." You are teaching them both planning skills and emotional awareness.

When they fall short, hold them accountable with empathy. This is where empathetic accountability becomes leadership in action. If they forget your birthday again or do the bare minimum, address it calmly. "I noticed you did not make any plans this year. I am hurt by this. What can you do differently next time?" Avoid harsh lectures or passive-aggressive silence. Stay composed and consistent. This teaches kids that relationships require effort and repair, not perfection or punishment. Do not brush it off with "It's fine" or explode with anger. Balance truth with grace.

Then nurture a growth mindset by reinforcing how learning to give thoughtfully is a process. No one is born naturally generous. People learn through repetition, reflection, and example. Praise effort even when the result is imperfect. "I loved that you thought to bake my favorite dessert. That meant a lot." or "I appreciate the note you wrote. It shows you were thinking about me." This approach turns awkward attempts into valuable practice. Each time they try, they build emotional intelligence and confidence.

You can also make generosity a shared family practice. Create a "gratitude circle" or "appreciation swap" where everyone takes turns recognizing another family member. It can be as simple as saying, "I want to thank you for helping with dinner," or "I appreciate how you made me laugh today." At first, teens may roll their eyes, but consistency transforms resistance into routine. You are cultivating empathy through repetition. Over time, this habit builds emotional maturity and mutual respect.

Stop buying your own gifts or handing them cash for your birthday. It feels easier in the moment but teaches them that giving to you is an obligation, not an expression of care. Tell them, "I will not give you money for my present this year. I want you to think about what would make me feel appreciated." It may feel awkward or uncomfortable, but it is essential. Let them struggle a little. The discomfort is where growth happens. When they fail, remind them to keep trying. They enjoy nice gifts and experiences, they must also learn to create them.

Show them what thoughtful giving looks like in practice. Talk about how you plan for their birthdays. Say, "I put thought into your special day by trying to imagine what would genuinely reflect your unique interests. My hope is that you'll carry that same level of caring consideration into your choices for others." You are demystifying the work behind generosity. You can even make it a shared activity by helping them plan a small celebration for someone else. These experiences teach both empathy and responsibility.

Use humor to diffuse tension. When you introduce these new expectations, teens might roll their eyes or accuse you of overreacting. Respond with warmth and wit. "I know, I know, you think I'm dramatic. But we can fix this before my next birthday, right?" Light humor keeps the atmosphere safe while signaling you expect change. Your tone communicates leadership grounded in patience rather than frustration.

AI can also help you maintain consistency. Ask your AI assistant to schedule reminders before important dates so you can start conversations early instead of waiting until emotions run high. Use it to generate discussion questions like, "What can we do to make birthdays feel special for everyone this year?" or to create shared family calendars that include birthdays, milestones, and gratitude days. By integrating technology intentionally, you turn it into a bridge for connection rather than a distraction from it.

Teaching your children generosity is not about money; it is about awareness. Encourage them to notice small opportunities to give such as helping with chores without being asked, complimenting someone, writing a thank-you note, or giving time and attention. Ask AI to help you come up with "family giving challenges," like spending a week doing one kind thing per day. Talk about what they learned at the end of the week. You are helping them build the habit of generosity through reflection and repetition.

Remember, children learn more from what they see than what they are told. If they rarely hear you celebrate your own accomplishments or watch you accept appreciation gracefully, they may assume adults do not need acknowledgment. Change that narrative. Talk about your wins at the dinner table. Say, "I had a hard day, but I'm proud of how I handled it." or "I finished a big project at work, and I feel good about it." Invite them to celebrate with you. You are showing them recognition is not vanity; it is gratitude in action.

At the same time, be mindful of over-giving. If you always go above and beyond for every celebration, your children may feel they can never match your effort. Simplify your approach. Choose thoughtful over extravagant. Say, "I am giving smaller gifts this year because I want to show thoughtfulness matters more than cost." This resets expectations and levels the emotional playing field.

When conversations about generosity get tough, stay calm and consistent. Teens are quick to label difficult discussions as "guilt trips." Avoid accusatory language like, "You never think about me." Instead, use language that focuses on growth and collaboration. "I was disappointed when my birthday was overlooked. I want us to do better together." You are inviting them to participate in the solution rather than defending against blame.

Establish rituals to make giving normal, not exceptional. Assign family members small roles in holiday or birthday preparations so everyone contributes. Rotate who chooses the meal, decorates, writes cards, or plans activities. This approach teaches planning, consideration, and shared joy. Use AI to help create checklists and reminders so no one forgets. The more structured the process, the easier it becomes for teens to participate successfully.

As you work through this, remember progress may come slowly. Your children may not transform overnight into empathetic gift-givers. They may stumble, forget, or get it wrong. But with emotional intelligence, clear expectations, consistent problem-solving, empathetic accountability, and a growth mindset, they will begin to change. When they make any effort at all, however small, acknowledge it. "Thank you for remembering. That means a lot." Recognition reinforces learning far more effectively than criticism.

Ultimately, modeling generosity in a selfie world is about teaching your children relationships are built on reciprocity, appreciation, and care. It is about reminding them love is not a one-way street. When you model empathy, structure accountability with kindness, and celebrate progress instead of perfection, you are not just raising children who remember birthdays. You are raising future adults who understand generosity is the language of love.

AI Prompts for Parents

Prompt 128 - Parent: Gift Giving Conversations

Actor: You are a Parenting Assistant

Background: I want to help my teen understand gift giving is about thoughtfulness, not price tags. My birthday is coming up, and I want to use it as a chance to talk about what meaningful giving looks like, how to express appreciation, choose gifts reflecting care, and set realistic expectations around celebrations. I want to guide my teen toward generosity, gratitude, and emotional intelligence without making the conversation feel like a lecture.

Call to Action: Create a list of warm, age appropriate conversation starters I can use to talk with my teen about my upcoming birthday. The goal is to encourage reflection about giving, gratitude, and the spirit of celebration in a way that strengthens our connection. Include a few questions that invite curiosity and empathy rather than guilt or pressure.

Data Check: If you need more information, ask me questions before proceeding.

Prompt 129 - Parent: Speaking from the Heart

Actor: You are a Family Communication Coach

Background: I have been feeling unappreciated and overlooked by my children. I give a lot of time, energy, and love to our family, yet lately it feels like my efforts go unnoticed. I want to communicate my feelings honestly, but also with grace and empathy, so my children understand how their words and actions affect me without feeling attacked or guilty. My goal is to open the door to mutual understanding and respect.

Call to Action: Help me create a clear and compassionate message that expresses my feelings of being unappreciated in a way that invites reflection and connection. I want my words to inspire awareness, gratitude, and a more balanced relationship rather than defensiveness or conflict.

Data Check: If you need more information, ask me questions before proceeding.

Prompt 130 - Parent: Celebrating Together

Actor: You are a Family Organization Assistant

Background: I want to plan a family celebration where everyone feels included and has a role. I would like to make the planning process simple, collaborative, and fun so no one feels overwhelmed or left out.

Call to Action: Create a clear, easy checklist that assigns tasks to each family member and keeps the celebration organized and enjoyable.

Data Check: If you need more information, ask me questions before proceeding.

AI Prompts for Young Leaders-in-Training

Prompt 131 - Student: Thoughtful Gift Ideas

Actor: You are a Personal Assistant

Background: I want to buy a thoughtful gift for my [enter here]. They enjoy [enter here], [enter here], and [enter here]. I would like something to match their interests and feel personal, not generic.

Call to Action: Suggest creative and meaningful gift ideas under $[enter here] that show care and attention to what they enjoy.

Data Check: If you need more information, ask me questions before proceeding.

Prompt 132 - Student: From the Heart

Actor: You are a Writing Coach

Background: I want to make a handmade birthday card for my [enter here]. They enjoy [enter here], [enter here], and [enter here]. I want the message to feel heartfelt and genuine while reflecting their personality and interests.

Call to Action: Provide me with reflection questions that will help me come up with sincere and personal language for the card, along with creative ideas to make it bright and meaningful.

Data Check: If you need more information, ask me questions before proceeding.

Prompt 133 - Student: Remembering What Matters

Actor: You are a Life Skills Coach

Background: I have trouble remembering birthdays and special occasions for people I care about. I want to be more thoughtful and organized so I can celebrate them on time and show I care. I am in the [enter here] grade and I am [enter here] years old.

Call to Action: Help me create an easy plan to remember birthdays and special events. Include ideas for setting reminders, keeping track of dates, and preparing simple, meaningful ways to celebrate.

Data Check: If you need more information, ask me questions before proceeding.

CHAPTER 22

Healing the Space Between Us

"Sometimes my children act so hateful toward me. I have made every sacrifice to the point of exhaustion, from trying to be the perfect parent. It feels impossible to meet their demands. How can I improve our relationship?"

I f this scenario feels like it could have come from your own journal, take a deep breath. You are not alone. Many parents reach a point where love starts to feel one-sided, where effort goes unnoticed, and where exhaustion feels like the only thing you are giving consistently. Feeling unappreciated by the people you love most is not a sign of failure; it is a sign of depletion. It means your emotional tank is running on fumes and your family's engine light is blinking "maintenance required." The good news is acknowledging burnout is not a betrayal of your love, it is the first act of healing.

Let's begin by challenging one of the most dangerous myths of modern parenting: Love is proven through sacrifice. We are told good parents give everything, their time, rest, dreams, patience, and ask for nothing in return. Self-erasure does not equal devotion. It teaches children caring for others means abandoning yourself, which is not love. It is martyrdom disguised as virtue. Leadership in the family begins when you decide love must include you too.

Your role as the leader of your home is not to give until you collapse. It is to model how to live well, love well, and rest well. If you want your children to grow into compassionate, self-regulated adults, they need to see what that looks like. Tell them the truth about your limits. Say, "I love being with you, but I am feeling tired right now. I need some time to rest so I can be my best self." This simple honesty is emotional intelligence in action. It shows your children caring for yourself is not selfish.

Children often push parents past their limits not out of malice but out of instinct. Their world revolves around their own needs. They are learning through your example where love begins and ends. When you never rest, they assume rest is optional. When you never say 'no', they assume boundaries are negotiable. Your actions are their manual for how relationships work. If you teach that love means constant availability, they will one day repeat the same pattern in their own relationships, either by overgiving or expecting others to do it for them.

Setting the emotional tone for your family starts with calm, transparent communication. Instead of bottling resentment or guilt, express your needs openly. You might say, "When I am working or resting, I need quiet. You can ask for help after I'm finished." You are not rejecting them; you are teaching them every person has limits that deserve respect. Define together what qualifies as "urgent" so boundaries feel clear, not arbitrary. Then practice enforcing those boundaries consistently. If they interrupt anyway, respond evenly: "I hear you. I will help when I'm done resting." Consistency is the quiet language of love.

Of course, enforcing boundaries will trigger discomfort, especially if your children are used to unlimited access. They may cry, pout, or lash out. You may feel guilty, hearing your inner voice whisper, "Good parents don't need breaks." What an untruth; constant availability does not make you more loving. It makes you depleted. A depleted parent cannot lead with clarity or compassion. Children deserve your calm presence more than your constant presence. If you need to take time out, take it!

When your child says something hurtful like "I hate you" or "You're the worst," resist the urge to match their intensity. Step into your leadership role by staying steady. Say, "I hear that you are upset. It is okay to feel angry, but it is not okay to speak that way. Let's talk when we are both calm." For older children, try, "I want to understand what is making you angry, but I won't stay in a conversation that turns disrespectful." You are modeling emotional control and accountability, showing that respect and love can coexist even in conflict.

Teaching problem solving in these moments means inviting your child to participate in repair. After an outburst, say, "We both got upset earlier. What can we do differently next time?" This moves the focus from punishment to partnership. It also helps them learn the power of reflection and responsibility. If they are not ready to talk, give them space. Then, when the time is right, revisit the conversation with curiosity rather than criticism.

AI can serve as a practical support in this process. If you struggle to find gentle words for setting limits, ask your AI for help crafting a message that is both firm and loving. Use AI to create visual family schedules that include your rest time, so your children can see it as a shared expectation rather than a sudden boundary. You might even involve them in creating it, asking AI for fun ways to name those moments, like "quiet recharge hour" or "family calm time." This keeps structure from feeling punitive.

Holding your children accountable with empathy is one of the hardest parts of leadership. They will test your resolve. They may push against every new boundary. When they do, remind yourself their resistance is a sign of adjustment, not rebellion. Say calmly, "I understand you want my attention right now. I am resting, and I will come find you when I am done." You are teaching patience and self-regulation through consistency. Over time, they will internalize and know you mean it when you say "no". Your "no" is not rejection, but stability.

Let your children see your humanity. If you lose your temper, own it. "I yelled earlier because I was tired and didn't take my quiet time. Next time, I will do better." Humility turns mistakes into lessons. You are showing them growth is not about perfection, but about repair. This is what nurturing a growth mindset looks like at home, admitting your flaws, learning from them, and moving forward with grace.

Problem solving also means examining the systems that keep you stuck in exhaustion. Ask yourself: Am I trying to do everything alone? Can I delegate or share responsibilities? Leadership is not about doing all the work; it is about building a team. If you have a partner, include them. If you are a single parent, consider family, friends, or community resources. Even small acts of support such as a neighbor carpooling, a friend picking up groceries, or a shared calendar can lift the burden. AI can also help you plan these supports by generating checklists, simplifying communication, and helping organize routines that preserve your energy.

Healing the space between you and your children also requires patience. Sometimes emotional distance is not defiance but a reflection of their own confusion or pain. They may not know how to reconnect with you. Instead of forcing closeness, build it slowly through small, positive moments (i.e., a shared meal, a walk, or a few minutes of laughter). Let affection grow naturally again. Connection does not return through grand gestures. It rebuilds through consistency, calm, and care.

Finally, remember that family estrangement or tension does not always mean either side is broken beyond repair. Sometimes distance grows because neither side has yet learned the skills of repair. True strength is not found in cutting off people who hurt us, but in learning how to mend relationships with courage and boundaries intact. Healing is not instant; it is a practice of returning to love while honoring your own limits.

You are not failing your children by needing rest or by asking for respect. You are showing them what balanced love looks like. When they see you care for yourself with honesty, set boundaries with clarity, solve conflicts with empathy, and recover from mistakes with humility, you are teaching them the kind of leadership that heals, not just relationships, but generations.

AI Prompts for Parents

Prompt 134 - Parent: Asking for Quiet Time

Actor: You are a Parenting Assistant

Background: I feel guilty when I tell my children I need quiet time.

Call to Action: Help me phrase this kindly but firmly so they understand without feeling rejected.

Data Check: If you need more information, ask me questions before proceeding

Prompt 135 - Parent: Restoring Emotional Balance

Actor: You are an Emotional Regulation Coach

Background: I'm exhausted and shut down emotionally, and then I feel unworthy of my kids' love.

Call to Action: Suggest ways to recognize my own needs and set realistic expectations without guilt.

Data Check: If you need more information, ask me questions before proceeding.

Prompt 136 - Parent: Peaceful Time Apart

Actor: You are a Family Organization Assistant

Background: I want to make alone time part of our family routine without constant fights about it.

Call to Action: Create a simple plan or schedule that sets clear expectations for everyone.

Data Check: If you need more information, ask me questions before proceeding.

AI Prompts for Young Leaders-in-Training

Prompt 137 - Student: Understanding Quiet Time

Actor: You are a Personal Assistant

Background: My parents say they need quiet time and I feel sad or mad about it. I am in the [enter here] grade and I am [enter here] years old.

Call to Action: Help me think of ways to stay calm and respect their space without feeling unloved.

Data Check: If you need more information, ask me questions before proceeding.

Prompt 138 - Student: Waiting with Patience

Actor: You are an Emotional Regulation Coach

Background: I get upset when my parents say "no" to playing with me right away. I am in the [enter here] grade and I am [enter here] years old. I want to understand how to manage my feelings when things do not happen on my timeline.

Call to Action: Suggest ways to calm down, handle my disappointment, and wait patiently while still feeling connected to my parents.

Data Check: If you need more information, ask me questions before proceeding.

Prompt 139 - Student: Planning Togetherness

Actor: You are a Time Management Assistant

Background: I want to plan a fun time with my parents that works for both of us. I am in the [enter here] grade and I am [enter here] years old.

Call to Action: Help me think of ideas and times to ask them when they are rested and ready.

Data Check: If you need more information, ask me questions before proceeding.

CHAPTER 23

Leading with Compassion, Living with Grace

"Does anybody get parenting right? It feels so impossible and I feel so inept. No matter how hard I try I just can't seem to get it right. Am I the only parent who feels like this?"

You are not alone. Let's start with the truth every parent needs to hear: No one gets parenting right. Anyone who says they do is not telling the full story. The idea of perfect parenting is an illusion that leaves loving parents feeling ashamed, exhausted, and uncertain. There is no single correct way to raise children because each child, each season, and each family dynamic is constantly changing. Perfection is not the goal. Presence is.

If you have ever gone to bed replaying every sharp word or missed opportunity, you are not failing. You are caring. The very fact you are worrying about being a good parent is proof you already are one. The concern in your heart is not weakness. It is love in action.

Parenting is the only leadership role in the world where the rules shift every time you start to understand them. The toddler who adored you yesterday may resist you today. The child who once shared every thought might mumble their response. Just when you think you have mastered one stage, your children change again. There is no fixed target to hit. The challenge is not to be perfect, but to adapt with patience, humility, and grace.

As the leader of your family, your most important role is setting the emotional tone. It does not mean pretending to be calm when you are frustrated or wearing a fake smile when you feel overwhelmed. It means modeling how to recover and repair. When you lose your temper or feel stretched too thin, come back to your child and say, "That was not my best moment. I am sorry. Let's try again." This one statement teaches more about love and strength than any lecture ever could. Your children do not need flawless parents. They need parents who show them how to recover with honesty and heart.

It is time to let go of the belief you are supposed to know exactly what you are doing. Most of us were handed babies with no instruction manual, no training, and no guarantee our instincts would always be right. Even the most experienced parents are learning on the job. Every family is its own experiment in love and patience. Books and experts can offer guidance, but they cannot account for your family's unique mix of personalities, histories, and dreams. Parenting is not a science. It is a living relationship that asks for curiosity more than certainty.

Setting clear expectations and boundaries does not require perfection. It means deciding what your family values and returning to those values even when it is hard. It means saying, "We treat each other with respect," even when tempers flare. It means holding the line on chores, sleep, or technology, not to control your children but to protect their well-being. Your consistency gives them stability. Leadership in parenting is not about getting every decision right. It is about staying grounded in what matters most.

Self-doubt is part of responsible parenting. It shows you care enough to reflect. The pilot who double-checks their instruments is not insecure, they are wise. The same applies to you. Second-guessing yourself means you are paying attention. It means you are thinking critically about how to lead your family with love.

Now, let's talk about guilt, that persistent companion who never seems to take a day off. Some guilt is healthy when it nudges you to make things right. But, most of the guilt parents carry is unnecessary. It is shame disguised as responsibility. It tells you being tired means you are lazy or taking a break means you are selfish. It whispers lies that keep you comparing your messy reality to other people's carefully edited highlights. The next time guilt shows up uninvited, pause and ask, "Is this helping me grow, or is it just me punishing me for being human?"

Your children are learning from how you handle your own mistakes. If you talk to yourself with kindness after you mess up, they will learn to do the same. If you spiral into shame and harsh self-talk, they will too. You can shift this pattern by naming your humanity out loud. Say, "This was not my best moment, but I am learning." Show them that mistakes are not the end of the story. They are opportunities to learn and try again.

Having clear expectations and boundaries with your children begins with setting them for yourself. Know what kind of home you want to create. Do you want peace or perfection? Connection or control? Decide that your goal is not to raise happy children every moment but to raise resilient ones who know how to handle disappointment and frustration with grace.

Decide that you will not avoid conflict but will model how to work through it with respect. These are far more meaningful goals than getting every moment "right."

Problem solving is where the magic happens. It means turning mistakes into lessons rather than punishments. When you snap at your child, own it. Say, "I did not handle that well. Let's talk about how we can do better next time." When they fall apart because they are frustrated, say, "I can see you are upset. Let's figure out a way to handle this together." These conversations teach them relationships are built on repair, not perfection.

Empathetic accountability is what holds your leadership steady. It means guiding your children firmly but kindly. It means saying, "I love you, but that behavior is not acceptable," instead of, "You are bad." It will teach them consequences are not punishments. They are opportunities to reflect, take responsibility, and grow. When accountability comes with empathy, children feel both loved and guided.

Parenting with a growth mindset means understanding that everyone in your family, including you, is still learning. You will lose your patience, break a promise, or forget an important date. None of that makes you unworthy of love. It makes you human. When you respond to those moments with honesty and humility, you are showing your children how to face their own imperfections with courage.

AI can be a useful instrument in this work. When you are unsure how to approach a difficult conversation, ask AI to help you write a message balancing firmness and kindness. When guilt starts to creep in, ask for affirmations that help you reframe your thoughts. You might even use AI to create bedtime reflection questions for your children such as, "What was one mistake you learned from today?" or, "What did you do that made someone else feel seen or loved?" These small tools can help keep your family focused on growth, not performance.

So here is the truth that will set you free. Perfect parents do not exist. There are only real parents, imperfect, loving, exhausted, and endlessly trying again. Your children do not need you to be perfect. They need you to be present, patient, and honest about what learning looks like. When they see you lead with compassion and live with grace, they will learn to do the same. And that is what parenting done well truly looks like.

AI Prompts for Parents

Prompt 140 - Parent: Progress Over Perfection

Actor: You are a Parenting Assistant

Background: I have been feeling like I am failing as a parent and that no matter what I do, it is never enough. I want to shift my perspective and remember that parenting is about progress, not perfection.

Call to Action: Help me reframe these negative thoughts and remind me what healthy, realistic expectations for parenting look like so I can lead with more grace and self-compassion.

Data Check: If you need more information, ask me questions before proceeding.

Prompt 141 - Parent: Grace After Mistakes

Actor: You are an Emotional Regulation Coach

Background: I feel overwhelmed with guilt whenever I make a parenting mistake. Even small missteps make me question my worth and ability as a parent. I want to learn how to acknowledge mistakes honestly without letting them define me.

Call to Action: Give me practical ways to accept responsibility with self-compassion, learn from my mistakes, and release guilt so I can lead my family with confidence and peace.

Data Check: If you need more information, ask me questions before proceeding

Prompt 142 - Parent: Teaching Through Mistakes

Actor: You are a Family Communication Assistant

Background: I want to have an open conversation with my children about the fact that parents make mistakes too. I want them to understand that being a family means learning and growing together, even when I do not get everything right. I hope to model honesty, humility, and emotional intelligence through the way I talk about my own missteps.

Call to Action: Help me craft an age-appropriate conversation that demonstrates accountability while also teaching my children that mistakes are opportunities for growth and connection.

Data Check: If you need more information, ask me questions before proceeding.

| AI Prompts for Young Leaders-in-Training

Prompt 143 - Student: When Parents Mess Up

Actor: You are a Personal Assistant

Background: I get upset at my parents when they make mistakes. They should know better. I am in the [enter here] grade and I am [enter here] years old.

Call to Action: Help me understand how to respond kindly and why it matters.

Data Check: If you need more information, ask me questions before proceeding.

Prompt 144 - Student: Learning From Mistakes

Actor: You are an Emotional Regulation Coach

Background: I feel really bad when I make mistakes or get in trouble at home. I am in the [enter here] grade and I am [enter here] years old.

Call to Action: Suggest ways to calm down and see mistakes as learning opportunities.

Data Check: If you need more information, ask me questions before proceeding.

Prompt 145 - Student: Helping at Home

Actor: You are a Time Management Assistant

Background: I want to help my family have calmer, happier days. I am in the [enter here] grade and I am [enter here] years old.

Call to Action: Help me plan small ways to help out or cooperate better at home.

Data Check: If you need more information, ask me questions before proceeding.

PART III
Living Leadership

When Knowing Better Becomes Doing Better

Every family reaches a point where theory meets reality - where "I want to be more patient" faces a slammed door or a weary sigh.

This is where learning becomes lived experience.

Until now, we've focused on the mindset of the modern parent leader setting the emotional tone, communicating with care, and modeling emotional intelligence. Now, we turn toward practice: the daily choices that define our leadership and shape what our children see.

Part 3 invites you to apply reflective leadership to real moments such as setting boundaries, repairing after conflict, modeling gratitude, asking for help, and leading with calm authority. Each chapter offers stories, strategies, and AI-powered prompts to help you turn awareness into action.

This is the work of showing up anyway not because it's easy, but because it matters. Every pause, apology, and patient response becomes a lesson in love and leadership.

Here, perfection steps aside and presence takes the lead.

Take a breath, you've already done the hard part: choosing to keep growing.

Now, let's put your leadership into practice.

CHAPTER 24

Showing Up Anyway

P arenting has always been the hardest leadership role any of us will ever hold. None of us truly understood this the day we held our babies for the first time. We might have imagined sleepless nights, tantrums in the grocery store, or hard questions about life, love, and loss. What we could not have anticipated was the constant complexity of raising other human beings while we ourselves were still learning, healing, and growing. There is no license for parenting. No final exam to prove you know how to guide a stubborn preschooler or a heartbroken teenager. There is only the daily commitment to show up, even when you feel uncertain, tired, or afraid you are not enough.

Throughout these pages, we have explored the unglamorous truth that parenting is not about controlling our children. It is about leading them. True leadership, especially within a family, means showing up with humility and integrity, even when you have made mistakes. It means remembering the goal is not perfection but presence. You are not trying to create perfect children but to guide real ones toward becoming kind, capable, and compassionate adults. Even if you could parent flawlessly, which no one can, it would not guarantee your children would never struggle. They are not copies of you. They are unique individuals who will make choices, face challenges, and sometimes break your heart. It is the price of loving people who are still becoming.

If you have seen yourself in these chapters, the parent who gives and gives but feels overlooked, the one who snaps in frustration and later feels crushed by guilt, the one who loves fiercely but just needs a moment alone to breathe, know this truth. You are not broken. You are not failing. You are learning and leading. The fact that you care enough to keep trying means you are already displaying one of the most powerful acts of love. You are showing up, consistently.

We have talked about how giving too much without setting limits can unintentionally teach children to take us for granted. We have discussed how paying for our own gifts or saying nothing when we are hurt can teach children our needs do not matter. We have examined the cycle of yelling and how it might feel justified in the moment, but it always leaves us and our children wounded. Yelling might create silence, but it does not create safety. It might create obedience, but it does not create respect. The good news is recognizing this pattern is not shameful. It is freeing. Once we see what is happening, we can make new choices.

Real leadership begins when we pause. When we recognize our triggers and choose empathy over reactivity. When we use our calm voice to enforce boundaries. When we guide our children toward accountability with compassion instead of fear. Anger itself is not the enemy. What matters is how we handle it. Emotional intelligence in parenting does not mean you never feel anger. It means you have learned to name it, manage it, and use it as information instead of a weapon.

We have also talked about what it means to want space. Many parents feel guilty for craving time alone. The endless demands, noise, and needs can feel overwhelming. Wanting quiet does not make you ungrateful. It makes you human. Setting healthy boundaries is not rejecting your children. It is teaching them relationships thrive when everyone has space to rest, recover, and return. By modeling balance, you show your children caring for others includes caring for yourself.

Healthy boundaries are not walls. They are the bridges needed to protect connection by preventing resentment. Self-care is not indulgence. It is a responsibility. Rest and reflection give you the clarity to parent from intention instead of exhaustion. A rested parent is a better leader. You teach your children emotional regulation not only by what you say but by how you treat yourself.

Parenting will always include moments of regret, frustration, and fatigue. Yet, every time you come back to the table, every time you choose calm over chaos, you are building trust. Presence, not perfection, is what shapes your children's understanding of love. They are watching how you handle disappointment, how you repair after conflict, and how you keep showing up even when things are hard. This is how they learn what love in action looks like.

You will make mistakes, and so will they. You will both grow. It is what leadership in families really means. It is not command and control. It is humility, consistency, and courage. It is the willingness to keep learning in real time.

AI Prompts for Parents

Prompt 146 - Parent: Showing Up Anyway

Actor: You are a psychologist specializing in working with parents.

Background: I tend to focus on what went wrong in my parenting day rather than what went right. I want to build more confidence and peace in my daily rhythm. I also need to learn how to replace perfectionism with presence.

Call to Action: Create a "Showing Up Anyway" nightly reflection journal with three prompts: one about connection, one about learning, and one about gratitude. Also, help me identify three affirmations to remind myself progress matters more than perfection.

Data Check: If you need more information, ask me questions before proceeding.

Prompt 147 - Parent: Repairing After Anger

Actor: You are a psychologist specializing in helping parents repair connections after losing their temper.

Background: I yelled during a stressful moment and now I feel guilty and disconnected. I want to model accountability and emotional regulation. My child is [enter here] years old.

Call to Action: Create a script for repairing trust after conflict that matches my child's age. Include a calm opener, acknowledgment of my behavior, and an invitation to talk. Provide me with two follow-up affirmations to repeat before the next hard conversation.

Data Check: If you need more information, ask me questions before proceeding.

Prompt 148 - Parent: Boundaries with Love

Actor: You are a psychologist specializing in teaching parents saying "no" can be an act of care.

Background: I often over-give and feel resentful or invisible. I want my children to see that love includes healthy limits.

Call to Action: Create a Boundary Communication Plan that includes: (1) a calm boundary-setting script, (2) language for follow-up when my kids push back, and (3) a short mantra to say when I feel guilty enforcing limits.

Data Check: If you need more information, ask me questions before proceeding.

Prompt 149 - Parent: Leading with Calm

Actor: You are a technologist specializing in teaching parents how to use AI as a tool, not a replacement for parenting.

Background: I often feel mentally overloaded and reactive. I want to pre-plan my tone and approach before challenges arise.

Call to Action: Generate calm-response scripts for my most stressful moments (morning rush, homework battles, bedtime routines). And provide a weekly emotional check-in template to help me track my triggers and wins.

Data Check: If you need more information, ask me questions before proceeding.

Prompt 150 - Parent: Family Gratitude Circle

Actor: You are a psychologist specializing in helping parents cultivate gratitude without guilt.

Background: My children are beginning to take gestures for granted, and I want to shift the family culture toward appreciation.

Call to Action: Create a Family Gratitude Circle plan with daily or weekly prompts, conversation starters, and a rotation for who leads. Help me develop three short affirmations for modeling appreciation authentically.

Data Check: If you need more information, ask me questions before proceeding.

Prompt 151 - Parents: Accepting Support Gracefully

Actor: You are a psychologist specializing in teaching parents asking for help is strength, not weakness.

Background: I've been exhausted, trying to manage everything alone. Help me learn how to model what healthy interdependence looks like.

Call to Action: Create a Self-Care Support Plan that includes (1) who or what can help me this week, (2) how I'll ask for help, and (3) a daily affirmation reinforcing that accepting support is an act of love.

Data Check: If you need more information, ask me questions before proceeding.

CHAPTER 25

Pulling It All Together

Parenting is the hardest leadership role any of us will ever hold. There is no certification, no final exam, and no corner office waiting at the end. No matter how many books we read, podcasts we have heard, or experts we followed, none of us ever received a neat diploma declaring us "Certified Parents Who Know What They Are Doing." What we did receive was a lifetime of uncertainty, second-guessing, messy conversations, and the daily reminder we were raising actual human beings who refused to follow any script we imagined for them.

It often felt impossible to get it right because the definition of "right" changed with every new season of our children's lives. One moment we were praised for being patient, the next we were told we were too soft. One expert encouraged us to give more freedom, another insisted we were creating chaos. Meanwhile, our children offered their own conflicting reviews, snuggling one day and slamming their doors the next because the sandwich was cut incorrectly.

Here is the truth that finally sets every parent free. Even if we could do everything perfectly, and we cannot, it still would not guarantee perfect children. It's not how human growth works.

Our children are not programs to be written or corrected. They are developing people with their own stories, emotions, dreams, and decisions. They will get things wrong. They will make choices that hurt or confuse us. They will disappoint and delight us, often within the same day. We are not failures or we have failed. It simply means we are raising real people who are still becoming.

Our job was never to make them flawless. Our job was to keep showing up; to be present even when we were weary or uncertain. We must keep trying even when we're angry or afraid. What our children learn most from us is how we respond when it is hard. When we stay in the conversation after conflict, when we own our mistakes, when we apologize sincerely and try again, we model the kind of resilience that sustains relationships.

For many of us, this kind of parenting was not modeled in our own childhoods. Some of us grew up in homes where authority ruled without question, where emotions were silenced or punished, and where mistakes carried shame instead of lessons. Even when our parents did the best they could, we often entered adulthood without the tools to repair relationships or regulate emotions. Learning new ways to lead our families meant teaching a language we were still learning ourselves. Of course it felt hard.

This book never promised a single secret or a perfect formula. There are no shortcuts to building emotionally intelligent families. Leadership in parenting is not about control. It is about intention. It is choosing how to respond even when everything in you wants to react. It is setting boundaries even when you would rather give in. It is modeling emotional control, empathy, and accountability so your children can one day lead themselves with wisdom and heart.

It is hard. It is daily. It is worth every effort.

In the chapter, "Parents Need Love Too," we explored the ache of being unappreciated. The parent who plans the birthday parties, buys the gifts, celebrates every win, and still ends up with a halfhearted text on their special day. It hurts. We have learned this disappointment often grows from patterns we helped create by over-giving, downplaying our needs, and avoiding honest conversations about appreciation. We have learned love is a two-way street. When we set clear expectations, teach gratitude, and invite our children to practice thoughtfulness, we shift the emotional tone of our homes from obligation to genuine care.

In "Screaming and Yelling," we faced one of the hardest truths in parenting: Yelling might feel powerful, but it does not create connection. We learned fear and respect are not the same. We explored how reactive communication damages trust and how calm consistency builds safety. This chapter helped us recognize our own triggers and gave us practical strategies for holding accountability with empathy. It reminded us that anger is not the problem; it is the unexamined reaction that causes harm. Emotional leadership means responding with composure, not perfection. It means showing our children big emotions can be managed without tearing relationships apart.

Then came "Unburning the Bridge" written for every parent who has ever felt too tired to give one more ounce of energy. The one who hides in the bathroom for five minutes of peace. The one who feels guilty for wanting silence. The one who believes needing space means they are failing. The "Unburning the Bridge" chapter invited us to reframe self-care as a form of family care. It reminded us boundaries protect love and rest restores compassion. We learned that taking time for ourselves models sustainability for our children; and, self-care is not selfish. It is responsible leadership.

Throughout each chapter, one truth echoed like a heartbeat: It is okay to ask for help. Parenting is hard even in ideal circumstances, and most of us do not live in ideal circumstances. When we feel anxious, angry, or hopeless, reaching out to a counselor, therapist, or trusted friend is not a weakness. It is wisdom. Asking for help means you care enough about your family to care for yourself. Your wellbeing is not separate from theirs. It is foundational. When you are supported, your children are safer.

We also learned to release the myth of perfection. We will forget permission slips, lose patience, and cave on screen time. We will say the wrong things, break promises, and compare ourselves to other parents. The measure of leadership is not in avoiding mistakes but in what we do next. When we return, apologize, repair, and grow, we teach our children how to do the same. They do not need parents who never stumble. They need parents who get back up.

So, we take a breath. We choose kindness toward ourselves. We remember that showing up, even imperfectly, is enough. We remind ourselves love grows stronger in the practice of patience. Leadership in families is not about mastery. It is about humility, presence, and grace.

If you need to reread this chapter every week, or every day, do it. Let it be a reminder you do not have to be perfect to be the parent your children need. Let it anchor you when guilt creeps in or exhaustion takes over. Parenting is not a performance. It is a relationship. It is a daily invitation to love, to learn, and to try again.

You are not alone in this. You are part of a generation of parents learning to lead with empathy instead of ego, with boundaries instead of burnout, and with courage instead of fear. You are raising the next generation of compassionate, confident, emotionally intelligent adults by being the kind of parent who keeps showing up.

That is leadership. That is love. And it is more than enough.

ABCD Framework Prompts for Reflection and Renewal

Prompt 152 - Parent: Leading with Calm Energy

Actor: You are an emotional wellness coach specializing in family leadership.

Background: I often respond to my children with frustration when I am tired or overwhelmed, and I want to set a calmer emotional tone at home.

Call to Action: Create a morning reflection routine to help me regulate my emotions before interacting with my children, including affirmations and grounding techniques.

Data Check: If you need more information, ask me questions before proceeding.

Prompt 153 - Parent: Family Meeting Blueprint

Actor: You are a family systems expert who helps parents communicate expectations effectively.

Background: My family often struggles with routines, and I feel frustrated repeating the same reminders. I want our home to feel structured but warm.

Call to Action: Develop a family meeting guide that helps us review weekly goals, schedules, and shared responsibilities with mutual respect and accountability.

Data Check: If you need more information, ask me questions before proceeding.

Prompt 154 - Parent: Accountability with Empathy

Actor: You are a parenting consultant who specializes in restorative communication.

Background: My children often shut down when I correct them. I want to hold them accountable in a way that builds trust rather than fear.

Call to Action: Create conversation scripts I can use when addressing mistakes so I can teach lessons with empathy and connection.

Data Check: If you need more information, ask me questions before proceeding.

Prompt 155 - Parent: Collaborative Problem Solving

Actor: You are a Leadership Strategist who helps families collaborate on everyday challenges.

Background: I want to model healthy problem solving by including my children in discussions about family rules, routines, and solutions.

Call to Action: Provide a step-by-step process for leading family problem-solving sessions that teach negotiation, compromise, and teamwork.

Data Check: If you need more information, ask me questions before proceeding.

Prompt 156 - Parent: Learning Through Mistakes

Actor: You are an Educator who helps parents and children build resilience together.

Background: My child tends to give up easily after making mistakes, and I sometimes struggle to stay patient when that happens.

Call to Action: Design a family reflection ritual we can do each week where everyone shares one mistake, one lesson, and one success to celebrate effort and growth.

Data Check: If you need more information, ask me questions before proceeding.

Prompt 157 - Parent: The 30-Day Leadership Reset

Actor: You are an executive coach who helps parents sustain balance and purpose.

Background: I want to maintain what I have learned in this book without losing focus over time. I need a plan that keeps me accountable to myself and my family.

Call to Action: Create a 30-day action plan that helps me practice emotional intelligence, consistent boundaries, and reflective leadership in my daily parenting.

Data Check: If you need more information, ask me questions before proceeding.

CHAPTER 26

Reflections

Parenting deserves the same kind of honest reflection great leaders give to their work. Leadership in any setting requires awareness, accountability, and courage. The same is true at home. Parents are leaders shaping the tone, rhythm, and culture of a family, often while managing exhaustion, uncertainty, and love so deep it hurts. Reflection allows you to pause long enough to make sense of it all to notice what is working, what is not, and what deserves more care and attention.

This is not about grading yourself or earning a perfect score. It is about checking in with the kind of parent and leader you are becoming. When you reflect, you are giving yourself permission to grow without guilt. You are saying, "I am still learning". The willingness to learn is the most powerful mindset any leader can hold. Reflection invites you to realign your actions with your purpose and your family's values. It helps you see how your choices shape the emotional climate of your home.

The truth is no one gets parenting "right." There is no certification, no graduation day, and no corner office waiting at the end. There is only the ongoing practice of showing up, paying attention, and trying again. Every time you pause to reflect, you strengthen your ability to lead with clarity instead of reacting out of habit. You learn to respond instead of explode, to guide instead of control, and to teach instead of shame.

These reflections are not meant to make you feel bad. They are here to help you see yourself more clearly. They are gentle mirrors, reminding you where you are strong and where you might want to shift. Take your time with them. Write down what comes up. Talk it through with your partner or a trusted friend. Share your insights with your children when they are ready. You are modeling self-awareness, humility, and growth, three of the most essential leadership skills your children will ever learn.

If reflecting feels awkward or unfamiliar, that is perfectly normal. Reflection is a muscle, and like any muscle, it grows stronger with use. You might start by asking simple questions: What emotional tone have I been setting at home lately? How do I respond when I am frustrated? Do I model calm problem-solving or reactive behavior? Am I showing my children what it looks like to set healthy boundaries, care for myself, and take responsibility when I am wrong?

Your answers will change over time, because parenting changes you. Again, the goal is not to achieve perfection but to remain aligned with your purpose. When you reflect, you reconnect with your "why", the values driving your parenting decisions. You remind yourself love is not just about giving but about guiding. Leadership at home is not about control but about creating safety, trust, and consistency.

Reflection is also how you turn awareness into action. This is where the ABCD Framework becomes your companion. After you notice what needs attention, translate your reflection into a prompt that helps you take the next step. Define your **Actor**, the voice or role you want the AI to play, such as a parenting coach, counselor, or communication expert. Offer a **Background,** give honest context about what is happening in your family or within yourself. Write a clear **Call to Action,** explain what you want to learn, build, or repair. Then close with a **Data Check**: "If you need more information, ask me questions before proceeding." This simple structure transforms reflection into progress. It turns vague thoughts into personalized, supportive guidance you can use right away.

Your values are the foundation of your leadership. They influence your tone, your decisions, and the way your family experiences love and structure. If you are not clear about them, you risk reacting instead of leading. Take time to name the values most important to you such as kindness, honesty, respect, consistency, or joy. Say them aloud. Write them down. Let your children hear you talk about them. When everyone understands the "why" behind your rules and routines, it becomes easier to follow them with trust instead of resistance.

Ask yourself what story you want your children to remember about how you led your family. Do you want them to remember a home where people felt safe enough to make mistakes and learn from them? A place where laughter was as common as rules? A space where everyone, including you, was allowed to rest, recover, and start over? Reflection helps you create a purposeful story.

You will not get everything right, and that is exactly how it is supposed to be. Growth is not a straight line. It loops, tumbles, and rises again. What matters is that you keep showing up. Keep pausing long enough to notice the lessons in front of you. Keep asking better questions. Keep turning those questions into action.

Because at the heart of it all, reflection is love in practice. It is saying, "I am willing to see myself clearly so I can love you better." It is the quiet strength that keeps a family grounded in grace, guided by intention, and led by a parent who understands leadership at home begins with self-awareness.

Values Reflection

Your values are the foundation of your parenting. They are the "why" behind your rules, your routines, and your reactions. If you never articulate them clearly to yourself or to your family, it's easy to lead reactively instead of intentionally.

Ask yourself:

1. What are the top 3-5 values for which I want my family to be known?

2. How did my own upbringing shape what I value now? Which lessons do I want to pass on? Which do I want to change?

3. When I think about the kind of adults I hope my children will become, what traits or values stand out?

4. Do I ever explain why our family values certain things to my children? Or, do I expect them to absorb it without discussion?

5. Where do I see a gap between what I say I value and how I actually parent day to day?

6. How do I model our values under stress or conflict?

7. How do I want my kids to treat others? How do I want them to treat themselves?

8. Where in our family life do our values show up most clearly? Where are they getting lost?

These questions are not about identifying all your flaws, they are about clarity. When you are clear on your values, you can make decisions more confidently, enforce boundaries more fairly, and communicate expectations without confusion.

Consider journaling your answers or discussing them with a co-parent or trusted friend.

Planning Idea:

Ask your AI assistant:

- "Help me write a family values statement based on these priorities."

- "Suggest ways to talk to kids about empathy/respect/responsibility in age-appropriate ways."

- "What activities help reinforce values like kindness or teamwork at home?

Prompt 158 - Parent: Self-Reflection and Awareness

Actor: You are a certified leadership coach who specializes in helping parents strengthen self-awareness and emotional regulation.

Background: I have been feeling reactive lately when my children argue or ignore directions. I want to understand the emotional patterns behind my reactions and how to respond with more calm and clarity.

Call to Action: Help me reflect on my emotional triggers and design a simple daily check-in routine that builds awareness before I respond. Include affirmations and reflection questions I can use each evening.

Data Check: If you need more information, ask me questions before proceeding.

Prompt 159 - Parent: Values Alignment

Actor: You are a family values strategist who helps parents articulate and communicate their guiding principles.

Background: I have a general sense that kindness, honesty, and consistency matter in our home, but I have never written them down or discussed them openly with my kids.

Call to Action: Guide me through a process to identify, define, and communicate our top five family values. Then help me write a short Family Values Statement I can post in our home and review with my children.

Data Check: If you need more information, ask me questions before proceeding.

Prompt 160 - Parent: Turning Reflection into Action

Actor: You are a parenting behavior specialist who teaches families how to turn self-reflection into actionable routines.

Background: I often realize what I want to change after I lose my patience, but I struggle to apply that insight the next time. I want to build habits that connect reflection to real behavior change.

Call to Action: Help me design a "Reflect-Respond-Repair" action plan I can use whenever conflict arises. Include simple language I can model for my children that shows accountability without shame.

Data Check: If you need more information, ask me questions before proceeding.

Prompt 161 - Parent: Modeling Growth for Children

Actor: You are a child development expert who helps parents demonstrate humility and growth mindset through everyday conversations.

Background: I want my children to see that adults can admit mistakes, learn, and start again. Sometimes I struggle to find the right words to share these lessons without overexplaining or sounding forced.

Call to Action: Provide conversation scripts and examples I can use to show my children what it looks like to take responsibility, apologize with sincerity, and express a commitment to improvement.

Data Check: If you need more information, ask me questions before proceeding.

Prompt 162 - Parent: Re-Centering During Stress

Actor: You are a mindfulness and family-wellness coach who integrates emotional regulation into leadership practice.

Background: When family life feels chaotic, I lose sight of my purpose and react instead of leading with intention. I want to practice pausing before speaking or deciding.

Call to Action: Create a three-step "Pause to Lead" exercise that helps me slow down, reconnect with my purpose, and respond from my family's core values during tense moments. Include a short breathing or grounding practice.

Data Check: If you need more information, ask me questions before proceeding.

Prompt 163 - Parent: Family Reflection Ritual

Actor: You are an AI family-leadership facilitator who designs reflective family discussions to strengthen communication and trust.

Background: I want to introduce a weekly family reflection time where we share what we learned, where we grew, and what we want to improve next week. I need help creating structure and prompts for it.

Call to Action: Design a 20-minute family reflection ritual with guiding questions for parents and children, a gratitude moment, and a closing affirmation that reinforces our values and connection.

Data Check: If you need more information, ask me questions before proceeding.

Routines Reflection

Routines are where values meet reality. They are the lived habits that shape your family culture, for better or worse. Reflecting on routines helps you see whether your daily life supports the family you want to build or undermines it in small, unnoticed ways.

Ask yourself:

1. Which parts of our day feel most stressful or chaotic? Why?

2. Where in our routines do we consistently argue or fight?

3. What parts of our day work well? Why do they feel smoother?

4. Do our routines make space for connection, or are we always rushing?

5. How much of our time is spent on screens, chores, homework, rest, play? Is this balance what I want?

6. Are our routines consistent enough for kids to know what to expect? Where is there confusion?

7. How much am I managing alone? Are there routines I could share or delegate?

8. Do our routines reflect our stated family values? For example, if I say family meals are important, do we make time for them?

9. How do I feel at the end of most days? Overwhelmed? Fulfilled? Disconnected? What might our routines have to do with that?

These reflections can be eye-opening without being blame-filled. No family routine is perfect. The point is to see clearly, so you can tweak intentionally.

AI Planning Partner Idea:

Ask your AI assistant:

- "Help me design a calmer bedtime routine for a 6-year-old."

- "Suggest ways to reduce morning chaos for school days."

- "Create a weekly schedule template that includes chores, homework, and family time."

Prompt 164 - Parent: Calming the Morning Routine

Actor: You are a family efficiency coach who helps parents create calm, connection-based morning routines.

Background: Our mornings often feel rushed and tense. Everyone scrambles to get ready, and small issues turn into arguments. I want our mornings to start with peace and purpose instead of stress.

Call to Action: Help me redesign our morning routine to include clear roles, preparation the night before, and at least one short moment of positive connection before leaving the house.

Data Check: If you need more information, ask me questions before proceeding.

Prompt 165 - Parent: Bedtime Reset

Actor: You are a parenting and sleep consultant who specializes in creating peaceful evening transitions.

Background: Bedtime feels chaotic. My children resist winding down, and I often lose patience trying to get them to bed. I want to end our days with calm, connection, and consistency.

Call to Action: Create a structured yet nurturing bedtime routine that helps children relax and feel emotionally secure. Include suggestions for lighting, language, and rituals that signal rest.

Data Check: If you need more information, ask me questions before proceeding.

Prompt 166 - Parent: Balancing Family Time and Screens

Actor: You are a digital wellness coach who helps families balance technology use with meaningful in-person connection.

Background: Our evenings are often dominated by screen time. I want to maintain digital balance without constant nagging or power struggles.

Call to Action: Design a realistic weekly screen-time plan that sets clear expectations, includes family tech-free zones, and offers fun alternative activities that align with our values.

Data Check: If you need more information, ask me questions before proceeding.

Prompt 167 - Parent: Shared Responsibility Routine

Actor: You are a family systems strategist who teaches parents how to share daily responsibilities fairly.

Background: I feel like I carry most of the household management like meals, laundry, and logistics. I want to model teamwork by delegating more and involving everyone appropriately.

Call to Action: Help me create a shared family responsibility chart that distributes daily and weekly tasks by age and ability. Include accountability check-ins that build consistency without nagging.

Data Check: If you need more information, ask me questions before proceeding.

Prompt 168 - Parent: Family Meal Connection

Actor: You are a family culture consultant who focuses on strengthening connection through everyday routines.

Background: We say family meals are important, but we rarely sit down together. I want to make shared meals a predictable part of our rhythm without adding pressure or guilt.

Call to Action: Help me design a simple weekly meal schedule that makes family dinners achievable at least three nights per week, including conversation starters and connection rituals.

Data Check: If you need more information, ask me questions before proceeding.

Prompt 169 - Parent: Weekly Rhythm Review

Actor: You are a leadership coach for parents who integrates reflection into family planning.

Background: Our routines shift quickly between school, work, and activities, and I rarely pause to see whether they reflect our values or just our obligations.

Call to Action: Create a Sunday reflection and planning template. I should help me review what worked, what felt stressful, and what needs adjustment for the coming week. Include questions to help reconnect me to our family's priorities.

Data Check: If you need more information, ask me questions before proceeding.

CHAPTER 27

Leadership Style Reflection

Your leadership style is the way you guide, teach, and set the emotional climate in your home. Unlike corporate leadership, it isn't just about strategy or policy, it's deeply personal and shows up in everyday moments: How you handle conflict, how you set boundaries, how you repair after mistakes. Reflecting on your style isn't about ranking yourself as "good" or "bad" but about getting curious about what you are modeling and what you want to model.

Ask yourself:

- When I think about "leadership" in my family, what words come to mind? Calm? Firm? Flexible? Reactive? Distant? Present?

- Do I lead mostly by giving orders, negotiating, collaborating, or avoiding conflict?

- How do I handle it when someone challenges my authority?

- Do I enforce rules consistently, or do I tend to give in to avoid meltdowns?

- When conflict happens, do I lean toward control, withdrawal, or connection?

- How do I repair a situation with my child after I lose my temper or make a mistake?

- Do my kids see me as safe to talk to, even about hard things?

- How do I respond when my child is emotional or upset? Do I try to fix it immediately, dismiss it, punish it, or listen?

- Am I modeling the emotional regulation I want them to learn?

- How would I want my child to describe my leadership when they are grown?

These aren't easy questions, but they are powerful. If you find yourself cringing at your own answers, that's not failure. That's the exact moment where growth begins. Leadership is learned. You can choose to lead differently tomorrow than you did today.

AI Planning Prompt:

Ask your AI assistant:

- "Help me phrase calm responses when my child talks back."

- "How can I explain rules without sounding harsh?"

- "Give me 5 ways to apologize sincerely to my child."

Prompt 170 - Parent: Identifying Your Leadership Style

Actor: You are a family leadership coach who helps parents identify and strengthen their natural leadership style.

Background: I want to understand how I currently lead my family. Sometimes I feel calm and consistent, but other times I'm reactive or avoid conflict. I want clarity about what leadership behaviors I'm modeling.

Call to Action: Help me identify my current leadership style using examples from family life. Then suggest three intentional adjustments that align my actions with the values I want to model for my children.

Data Check: If you need more information, ask me questions before proceeding.

Prompt 171 - Parent: Repairing After Mistakes

Actor: You are a parent-child communications expert who teaches emotional repair and accountability.

Background: When I lose my temper or make a mistake, I sometimes struggle with what to say to my child afterward. I want to repair without shame and show them that leaders take responsibility.

Call to Action: Provide me with a step-by-step "repair conversation" script modeling humility, accountability, and reassurance. Include affirmations I can use to rebuild trust after conflict.

Data Check: If you need more information, ask me questions before proceeding.

Prompt 172 - Parent: Leading with Calm Authority

Actor: You are a family leadership mentor who helps parents balance calmness with firmness.

Background: I often feel torn between being too strict and too lenient. When my child talks back or challenges my authority, I either overreact or withdraw.

Call to Action: Coach me through three ways to respond to defiance that maintain calm authority while preserving connection. Include short example phrases and body-language cues.

Data Check: If you need more information, ask me questions before proceeding.

Prompt 173 - Parent: Modeling Emotional Regulation

Actor: You are a mindfulness and parenting specialist who helps families build emotional awareness.

Background: I want to model emotional regulation, but when my children are upset, I often jump to fixing or disciplining instead of listening. I want to handle those moments differently.

Call to Action: Create a reflective practice that teaches me how to pause, validate emotions, and respond with empathy before problem-solving. Include a few phrases I can use during emotional moments.

Data Check: If you need more information, ask me questions before proceeding.

Prompt 174 - Parent: Rebuilding Connection After Conflict

Actor: You are a family relationship therapist who helps parents and children reconnect after arguments.

Background: After conflict, things feel tense in our home. I want to help everyone feel safe and seen again without ignoring what happened.

Call to Action: Design a short "family repair ritual" I can use after conflict, something that includes reflection, reconnection, and a shared statement of understanding.

Data Check: If you need more information, ask me questions before proceeding.

Prompt 175 - Parent: Future Leadership Vision

Actor: You are a reflective parenting coach who helps families create a vision for the kind of leadership they want to embody.

Background: I want to think long-term about what my children will remember about my leadership. I want to define the words and qualities I hope describe me as their parent when they are grown.

Call to Action: Guide me through a journaling or visioning exercise that helps me write a short "Parent Leadership Mission Statement" based on how I want to lead my family.

Data Check: If you need more information, ask me questions before proceeding.

CHAPTER 28

Turning Reflection Into Action

Reflection alone is useful, but change happens when you turn insight into small, repeatable steps. That's why this chapter isn't just about self-audit questions, it's also about building your own personalized parenting plan for ongoing growth.

Start by choosing one small area to focus on at a time. It's tempting to overhaul everything at once, but that's a recipe for burnout and resentment. Instead, pick the theme that felt most charged when you were reflecting. Maybe it was realizing your emotional tone sets everyone on edge. Maybe it was seeing that your rules are unclear. Maybe you noticed you are exhausted from managing routines alone.

Once you pick your focus, ask yourself:

1. What does "better" look like in this area? Be specific.

2. What would a single small improvement be?

3. What's realistic for our family right now, not ideal, but doable?

4. What help do I need to make that change? From my partner? My kids? Friends? Professionals? Technology?

5. How will I know if it's working? What signs will I look for?

Let's say you realized your family mornings are chaos. Instead of declaring "We will never have stressful mornings again," you might set a goal like, "I want mornings to feel 10% calmer." It might mean setting out clothes the night before, prepping breakfast in advance, waking up 10 minutes earlier, or using a shared family checklist.

Or, maybe you realized your leadership style is reactive. Instead of promising never to yell again, your goal might be, "I want to notice when I'm about to yell and pause first." It might involve writing calming phrases on a note by the fridge, practicing breathing exercises, or even telling your child, "I need a minute so I can talk kindly." These are small but meaningful shifts. Parenting isn't changed by dramatic declarations but by tiny adjustments made consistently.

AI Planning Idea:

- "Suggest 5 ways to make school mornings calmer."

- "Help me design a bedtime routine that feels more connected."

- "Give me a simple breathing exercise to use when I'm angry."

- "Write me a reminder script for talking calmly about rules."

Using technology in this case isn't about outsourcing your parenting but supporting it. Think of your AI assistant as the teammate who never gets tired of helping you brainstorm or rehearse.

Prompt 176 - Parent: Choosing One Focus Area

Actor: You are a leadership coach who helps parents create realistic growth goals without overwhelm.

Background: I've reflected on several areas of family life that could improve, but I tend to take on too much at once and lose momentum.

Call to Action: Help me identify one small, high-impact focus area to work on first. Create a 7-day micro-goal plan with one clear action step each day to help me practice consistency.

Data Check: If you need more information, ask me questions before proceeding.

Prompt 177 - Parent: Reducing Morning Chaos

Actor: You are a family organization strategist who specializes in building calm, efficient daily routines.

Background: Our mornings are rushed and stressful, which sets a negative tone for the day. I want to make mornings calmer without expecting perfection.

Call to Action: Design a "10% Calmer Morning Plan" with small adjustments we can sustain, such as evening preparation, role assignments, and mindset reminders. Include a one-sentence daily affirmation I can post by the door.

Data Check: If you need more information, ask me questions before proceeding.

Prompt 178 - Parent: Practicing Calm Communication

Actor: You are a communication and emotional regulation coach who helps parents lead with calm authority.

Background: I often react quickly when I feel disrespected or ignored. I want to notice my triggers earlier and respond with calm instead of control.

Call to Action: Create a three-part "Pause Before You Speak" plan that includes one breathing exercise, one grounding phrase, and one gentle script to use when setting boundaries.

Data Check: If you need more information, ask me questions before proceeding.

Prompt 179 - Parent: Delegating and Sharing Responsibility

Actor: You are a family systems consultant who teaches parents how to share daily responsibilities effectively.

Background: I handle most household management alone and often feel exhausted. I want to build routines where my partner or children take ownership of certain tasks.

Call to Action: Help me design a shared responsibility chart with clear roles, time expectations, and check-ins.

Include strategies to introduce it positively so it feels collaborative rather than controlling.

Data Check: If you need more information, ask me questions before proceeding.

Prompt 180 - Parent: Turning Insight Into Habit

Actor: You are a behavioral change expert who helps parents turn insights into sustainable daily habits.

Background: I often reflect and set goals, but I struggle to follow through when life gets busy.

Call to Action: Build a 21-day "Small Steps Plan" to transform one reflection into an automatic habit. Include a reward or reflection ritual to celebrate progress each week.

Data Check: If you need more information, ask me questions before proceeding.

Prompt 181 - Parent: Tracking Progress with Intention

Actor: You are an accountability coach who specializes in reflective growth tracking for parents.

Background: I want a way to see my growth without judging myself harshly. I need a simple system that measures progress in behavior and emotional tone, not perfection.

Call to Action: Create a weekly reflection tracker that helps me log small wins, moments of calm, and improvements in family connection. Include guiding questions for Sunday review.

Data Check: If you need more information, ask me questions before proceeding.

Integrating Family Goals

Every strong leader knows goals are powerful, but shared goals are transformational. Once you have reflected and chosen one small change to focus on, the next step is to invite your family into the process. This is not about delivering a lecture or running a board meeting with an agenda and PowerPoint slides, although if it is your style, I fully support your commitment to excellence. This is about conversation, collaboration, and connection. Family leadership is relational. It is about leading with empathy, consistency, and courage. When you include your family in your thought process, you model the kind of transparency that turns change into teamwork.

Children can sense when something feels off. They notice the tension in your voice when mornings are rushed or the way you sigh when rules keep shifting. Pretending everything is fine only confuses them. Instead, invite them into your growth. You might say, "I have been feeling like our mornings are too stressful. I want us to work together to make them calmer." Or, "I do not like that I have been yelling so much. I am going to try to pause and speak more kindly." Or even, "I want us to have clearer rules about screen time so it does not turn into arguments every day." When you speak with this kind of honesty, you are showing your children what emotional intelligence looks like in action. You are teaching them leadership means owning your behavior, adjusting your approach, and trying again. It is one of the most powerful lessons you can model.

Once you have shared what you are working on, turn the conversation into collaboration. Ask your children what they think would help, how they would like things to feel, and what they might change if they could. Then listen without judgment. You may be surprised by what they say. Even very young children often have thoughtful and practical ideas. A six year old might suggest picking out clothes the night before. A teenager might say mornings would go better if everyone knew the plan in advance. These insights may seem small, but they reveal how your children experience family life and how eager they are to help make it better when they feel heard.

The real magic happens when you turn these shared reflections into action. You can create a simple family goal together that feels realistic and encouraging. It might sound like, "We want mornings to feel ten percent calmer," or, "We will eat dinner together three nights a week without phones," or, "We will use a pause plan when we feel frustrated." Write it down somewhere visible and revisit it regularly. This transforms your reflection into a shared family agreement, one built not on control but on commitment and care. It becomes a living document of how your family is growing together.

Like any leadership team, your family will need accountability, but accountability at home should always be wrapped in grace. The goal is not perfection; it is progress. Notice the effort as much as the outcome. When your child takes initiative, acknowledge it. When you remember to take a breath before raising your voice, celebrate that moment too. Say, "I appreciated how you got ready on time," or, "This was a calm morning, thank you for helping," or, "You reminded me to breathe. I appreciate you." These small affirmations reinforce everyone's contributions.

Integrating family goals is not about tightening control; it is about loosening fear. It is about showing your children that leadership at home is a shared responsibility, not a solo performance. Families thrive when they know what they are working toward and feel their voices shape the path. Every conversation like this reminds your children love is not passive; it is active, intentional, and ever evolving. When you integrate your goals with theirs, you transform reflection into momentum, and your home becomes not just a place of routines but a space of shared purpose and steady growth. This is the art of modern family leadership, one small and honest conversation at a time.

Modeling Problem-Solving and Accountability

One of the most important aspects of parenting leadership is showing your children what to do when things go wrong. It's easy to imagine leadership as only about setting rules and expectations but the real heart of it is how you respond to conflict, mistakes, and disagreements.

Conflict is not a sign something is broken in your family. It's a sign your family is full of real, complex human beings with different needs and feelings. Your goal isn't to eliminate conflict, it's to handle it and build trust in the process..

Ask yourself:

1. How do I usually respond when my child challenges or disagrees with me?

2. Do I feel comfortable letting my child express anger, sadness, frustration or do I shut those feelings down?

3. When I'm angry, do I model respectful conflict? Or do I escalate things with yelling, sarcasm, or threats?

4. How do I repair the relationship with my child after a conflict? Should I apologize? Should I explain why I got upset?

5. When my child apologizes, how do I respond? Do I make it safe to admit mistakes?

6. Do my kids know what accountability means? How do we practice it at home?

7. Are consequences in our home fair, consistent, and connected to behavior? Or are they punitive or arbitrary?

8. Do I teach problem-solving skills, or do I just dictate solutions?

9. When disagreements happen between siblings, how do I guide them to resolution?

Prompt 182 - Parent: Creating a Shared Family Goal

Actor: You are a family leadership coach who helps parents and children work together to set shared goals that strengthen cooperation and trust.

Background: I have reflected on areas I would like to improve, and I want to include my family in setting a shared goal that feels realistic and encouraging.

Call to Action: Help me design a family goal-setting conversation that includes three simple questions for everyone to answer, a sample script for introducing the discussion, and a visual template for writing our family goal.

Data Check: If you need more information, ask me questions before proceeding.

Prompt 183 - Parent: Guiding a Family Conversation About Change

Actor: You are a communication coach who helps parents lead open, non-judgmental family conversations.

Background: I want to talk to my family about a change we need to make, but I want to avoid sounding critical or controlling.

Call to Action: Provide me with a short conversation guide that opens the discussion with empathy, invites input from each family member, and ends with an action everyone agrees on. Include examples of questions that promote collaboration.

Data Check: If you need more information, ask me questions before proceeding.

Prompt 184 - Parent: Encouraging Children's Input

Actor: You are a child development specialist who helps parents invite children's perspectives in family decision making.

Background: I want to give my children a voice in shaping our routines and goals, but I am unsure how to guide the discussion so it stays respectful and productive.

Call to Action: Create a list of age-appropriate questions that help children share their thoughts about family routines, stress points, and improvements they would like to see. Include tips on how to respond with validation instead of correction.

Data Check: If you need more information, ask me questions before proceeding.

Prompt 185 - Parent: Building Accountability with Grace

Actor: You are a family systems coach who teaches gentle accountability and progress tracking in the home.

Background: Our family sets goals but often loses motivation after the first few days. I want a way to track progress without pressure or guilt.

Call to Action: Design a simple weekly accountability practice that celebrates small wins, includes reflection prompts, and helps us stay encouraged even when we miss a step.

Data Check: If you need more information, ask me questions before proceeding.

Prompt 186 - Parent: Modeling Growth and Humility

Actor: You are a parenting mentor who helps adults model emotional maturity and learning in front of their children.

Background: I want my children to see that I am learning too. When I make a mistake or forget our plan, I want to show them how to take responsibility without shame.

Call to Action: Write a few short phrases or scripts I can use when I need to admit a mistake and reset the tone in our home. Include examples of how to turn the moment into a family learning opportunity.

Data Check: If you need more information, ask me questions before proceeding.

Prompt 187 - Parent: Creating a Family Check-In Ritual

Actor: You are a family culture designer who helps parents build connections through simple weekly rituals.

Background: I want to create a weekly check-in where our family can reflect on what worked, what was stressful, and what we are proud of. I need it to be short and meaningful.

Call to Action: Create a 15-minute family check-in outline that includes an opening gratitude question, one reflection question for each person, and a closing affirmation we can say together.

Data Check: If you need more information, ask me questions before proceeding.

These questions help reveal the emotional patterns you are teaching. Your child isn't just learning what the rule is, they are learning how to disagree with someone respectfully, how to admit they were wrong without shame, how to listen and compromise.

Conflict is an incredible opportunity to teach emotional intelligence. But it can also reinforce fear and resentment if it's always about power or punishment.

AI Planning Idea:

- "Help me write an apology to my child after yelling."

- "Suggest ways to explain consequences calmly."

- "Give me phrases to use when mediating sibling fights."
- "How can I help my child brainstorm solutions instead of just telling them what to do?"

CHAPTER 29

Nurturing a Growth Mindset

P arenting invites us into the most meaningful leadership role we will ever hold. It also exposes how quickly we can slip into perfectionism. We want to get it right. We want our children to get it right. We imagine there must be a perfect way to discipline, teach, or guide that prevents pain and guarantees success. But leadership in a family is not measured by perfection, it is measured by growth. A growth mindset reminds us mistakes are not failures. They are evidence we are learning, stretching, and practicing what it means to be human together.

When you lead your home with awareness, you begin to see every challenge has something to teach. The spilled milk, the forgotten homework, the sibling argument, the parent who snaps at the end of a long day, all of it becomes a classroom. A home led with awareness is not one without frustration or conflict. It is one where emotions are noticed before they erupt, and where the leader, the parent, chooses calm over chaos. Emotional intelligence before action is what allows the family to recover quickly from tense moments. A parent who pauses to breathe instead of reacting teaches a lesson that words could never deliver. When you respond with empathy, you model maturity. You show your children that leadership does not mean controlling others; it means controlling yourself.

The emotional tone you set becomes the soundtrack of your home. Children learn from your mood even more than your instructions. If you bring humor to hard moments, they learn joy is a choice. If you speak gently when something breaks, they learn that mistakes can be repaired. If you demonstrate curiosity instead of anger when a rule is broken, they learn discovery is always possible. The tone you bring into each day communicates your values far more effectively than any rule ever could.

Consistency helps that tone take root. Families thrive when expectations are clear and boundaries are fair. A growth mindset does not erase structure; it deepens it. Rules rooted in fairness become lessons in accountability, not punishment. A child who understands the "why" behind the expectation learns to respect the process, not just the outcome. For example, a boundary such as "We speak kindly even when upset" teaches self-regulation. It does not just prohibit yelling; it invites awareness. Structure anchored in empathy is what gives freedom meaning.

Accountability can flourish only when it is grounded in empathy. Every leader must find the balance between firm guidance and grace. When your child disobeys or struggles, the goal is not to prove authority but to teach responsibility. Instead of asking, "Why did you do that?" you might ask, "What were you hoping would happen?" or "What could you do differently next time?" These questions shift the focus from guilt to growth. They build trust instead of fear. Children who experience compassionate accountability begin to develop inner discipline, the kind that lasts long after they leave your home.

Leadership at home is not about hierarchy; it is about modeling. Growth mindset leadership shows your family learning never ends. Let your children see you trying something new, asking for help, or recovering after a mistake. When they watch you handle your own frustration with honesty, they learn courage is quieter than perfection. You might say, "I am learning how to manage my time better," or "I was impatient earlier and I am working on handling it differently next time." These admissions do not weaken your authority; they strengthen it. They show your children you can be vulnerable and admit adults must sometimes acknowledge their imperfections.

Perfectionism can easily masquerade as high standards. Many parents push themselves so hard in the name of excellence that they forget that growth requires imperfection. You cannot teach resilience if you never model recovery. You cannot teach curiosity if you always have the answer. Growth happens when you release the myth that mistakes mean failure. A home that celebrates progress instead of perfection becomes a workshop for possibility.

Pay attention to how you speak about effort. If you often say, "You are so smart," try shifting to, "You worked really hard on that." When you praise strategy and persistence, your child learns to value process. The same is true for adults. When you say, "I learned something important today," or "That did not go how I hoped, but I see what to adjust," you remind yourself and your family that leadership is an ongoing experiment.

Families who practice a growth mindset learn failure is not the end of the story; it is the middle of learning. They develop language that celebrates trying, reframes setbacks, and honors the courage to continue. When your child brings home a disappointing grade, ask, "What did this show you about what you might do next time?" When you are frustrated by your own behavior, say aloud, "I am learning patience," or "That was tough, and I am giving myself grace." These statements become family affirmations that transform pressure into purpose.

You can cultivate this mindset intentionally. Create moments that celebrate effort instead of perfection. Start a family "learning wall" where everyone posts something they tried this week, successful or not. During dinner, ask, "What did you learn today that surprised you?" or "What challenge are you proud you faced?" These rituals normalize imperfection and turn growth into a shared experience.

Over time, you will begin to see a benefit nurturing a growth mindset transforms more than behavior, it transforms identity. Your children start to see themselves as capable of learning through struggle. You begin to see yourself as a leader who grows through reflection. The home becomes a place where everyone is free to make progress, not just meet expectations. This is the culture of thriving families: People who value curiosity over certainty, perseverance over perfection, and empathy over ego.

When you reflect on these ideas, notice the themes that call to you. Maybe you realize that you step in too quickly to fix problems instead of letting your child experiment. Maybe you see that you encourage academics but rarely praise emotional growth. Choose one focus area at a time. Change sticks when it is small, specific, and repeated. Ask yourself what would make that area ten percent better this month. Then take one step. The beauty of growth is that it compounds. Small adjustments practiced consistently build lasting transformation.

A growth mindset does not promise an easy life. It promises a purposeful one. It invites you to view every challenge as a chance to lead with intention. When you parent this way, you build a culture of resilience where everyone, including you, is free to evolve. Growth becomes your family language. And when your children describe you one day, they will not say you were perfect. They will say you were brave enough to keep growing.

AI Prompts in the ABCD Framework

Prompt 188 - Parent: Creating Family Language for Growth

Actor: You are a communication strategist who helps families design uplifting language that reinforces learning and resilience.

Background: I want our family to talk about mistakes, effort, and progress in a way that feels natural and positive.

Call to Action: Help me create a set of ten family phrases or affirmations to promote a growth mindset and replace critical or perfectionist language.

Data Check: If you need more information, ask me questions before proceeding.

Prompt 189 - Parent: Modeling Recovery After Mistakes

Actor: You are a family therapist who guides parents in demonstrating calm recovery after stressful moments.

Background: I want to show my children how to handle mistakes with grace rather than frustration or guilt.

Call to Action: Write a three-step reflection plan I can model aloud after I lose patience, including what to say, how to repair, and how to reset.

Data Check: If you need more information, ask me questions before proceeding.

Prompt 190 - Parent: Encouraging Resilience Through Reflection

Actor: You are a parenting coach who helps families strengthen confidence through reflection.

Background: My children become discouraged when they fail or fall short. I want to help them view setbacks as stepping stones.

Call to Action: Provide a short family reflection exercise that helps us talk about challenges with curiosity instead of judgment.

Data Check: If you need more information, ask me questions before proceeding.

Prompt 191 - Parent: Shifting from Control to Collaboration

Actor: You are a leadership consultant who specializes in helping parents guide rather than control.

Background: I want to move away from over-directing my children and invite them to help create solutions.

Call to Action: Help me design a collaborative problem-solving routine that balances my authority with my children's voice. Include prompts I can use to guide the discussion constructively.

Data Check: If you need more information, ask me questions before proceeding.

Prompt 192 - Parent: Building Confidence Through Celebration

Actor: You are a family engagement coach who designs rituals that honor progress and persistence.

Background: I want my family to celebrate small victories consistently without turning them into rewards or competition.

Call to Action: Create a monthly celebration plan including reflection questions, gratitude statements, and one shared affirmation we can repeat together.

Data Check: If you need more information, ask me questions before proceeding.

Prompt 193 - Parent: Teaching Emotional Intelligence Through Storytelling

Actor: You are an educator who uses storytelling to teach emotional awareness and self-regulation.

Background: I want to use stories from books, movies, or real life to help my children understand emotions and resilience.

Call to Action: Suggest five age-appropriate stories or storytelling activities that illustrate perseverance, empathy, and the value of learning from mistakes.

Data Check: If you need more information, ask me questions before proceeding.

CHAPTER 30

Pulling Reflections Into Daily Life

B y now, you have realized none of these questions has a single right answer. Leadership in a family is never about having all the answers; it is about continuing to ask better questions. Reflection is not a one-time event. It is a practice. Every stage of parenting invites new lessons, new challenges, and new growth. The goal is not to complete a checklist of insights but to cultivate an ongoing rhythm of awareness that strengthens how you lead your family each day.

Emotional intelligence before action is what turns reflection into progress. When you make time to notice how you feel before reacting, you transform ordinary moments into learning opportunities. A pause to breathe after a hard conversation, a moment of quiet before setting a new rule, or a brief journal note about something that triggered you, all of these are small acts of leadership. They remind you clarity comes before correction. You cannot guide your family well if you have not first checked in with your own emotional temperature. Leading with awareness is how you model composure, accountability, and empathy all at once.

The emotional tone you set through reflection is what holds your family together in unpredictable seasons. When you treat reflection as a way to recalibrate rather than criticize, your home becomes a space of calm inquiry instead of anxious reaction. Families thrive when leaders are steady and transparent. You might say, "I am realizing that we have been rushing through evenings and it feels too tense. I want to slow us down." This simple sentence can shift an entire week because it signals openness instead of control. Tone builds trust, and trust is what keeps connection strong during change.

The most effective families balance reflection with clear expectations and boundaries. A leader who regularly checks in with their values and routines is less likely to parent reactively. Taking time to ask, "Are our rules still serving our goals?" or "Is our schedule supporting our peace?" prevents resentment from building in silence. Boundaries built from reflection are strong because they are informed by purpose, not habit. This is where reflection meets strategy. It becomes less about fixing problems and more about aligning with what matters most.

Empathetic accountability is what keeps reflection grounded in compassion. When you revisit your choices, you are not grading yourself or assigning blame. You are learning. Reflection helps you see patterns without judgment and identify opportunities without shame. Maybe you notice that you raised your voice too often this month, or that you forgot to make time for fun. Accountability with empathy says, "I see where I can grow, and I am willing to keep practicing." This mindset frees you from guilt and turns self-awareness into momentum. It also teaches your children accountability is not punishment, it is progress.

Growth mindset leadership makes reflection an active part of daily life. Every great leader has a system for reviewing and adjusting direction, and family leadership is no different. Reflection keeps you adaptable. It helps you adjust to your child's changing needs, to your own evolving priorities, and to life's inevitable surprises. The family that reflects together grows together. Taking time to ask, "How are we doing?", creates shared ownership of your family culture. Your children learn leadership is not about pretending to be perfect; it is about staying curious and willing to change.

One of the simplest ways to make reflection real is through a regular check-in routine. You do not need elaborate systems or color-coded charts. All you need is ten minutes of honesty. Once a month, sit down with your journal or your AI assistant and revisit a few questions. Ask yourself what went well, what felt hard, and what you want to try again. If you have a partner, take a few quiet minutes together after the children are asleep. Ask, "How are we doing as a team? What's working? Where are we stretching thin?" These conversations deepen understanding more than any parenting book ever could because they invite transparency and collaboration.

You might use questions such as:

- What went well in our family this month, and what felt meaningful?

- Where did we struggle, and what might help next time?

- Did I live our family values clearly? Where did I fall short?

- How did I handle conflict? Did I model repair and responsibility?

- What is one small thing I want to do differently next month?

- What do I want to appreciate about my child and about myself?

These questions become anchors. They center you when life feels hurried and remind you leadership is a daily practice, not a performance. Ten minutes of honest reflection can change how you lead for the next ten days. You begin to respond with more thought, more kindness, and more intention because you have already done the inner work to see what matters most.

Reflection is how leaders stay aligned with their purpose. Without it, even the most loving families can drift into autopilot, reacting to schedules instead of shaping them. With it, you create a feedback loop between your values and your actions. Over time, reflection becomes a reflex, an instinct to pause, think, and realign before moving forward. The family that learns to reflect together learns to adapt without losing connection.

When reflection becomes part of your daily life, your family culture gains depth. Children feel safe because they know their home is guided by understanding, not impulse. You feel centered because you have a process for handling the messiness that comes with growth. Reflection transforms parenting from survival into leadership. It keeps your family evolving, connected, and grounded in grace.

AI Prompts in the ABCD Framework

Prompt 194 - Parent: Building a Monthly Family Check-In

Actor: You are a family systems coach who helps parents design meaningful monthly reflection practices.

Background: I want to start a family check-in routine that helps us pause, review, and realign with our values without it feeling like a meeting.

Call to Action: Create a ten-minute monthly check-in structure that includes guiding questions, a short gratitude practice, and one shared commitment for the next month.

Data Check: If you need more information, ask me questions before proceeding.

Prompt 195 - Parent: Strengthening Emotional Awareness Through Reflection

Actor: You are a parenting mentor who helps families increase emotional intelligence through simple daily habits.

Background: I want to be more aware of my emotional state before reacting to my children.

Call to Action: Design a three-step daily reflection practice that helps me notice, name, and navigate my emotions before engaging in family interactions.

Data Check: If you need more information, ask me questions before proceeding.

Prompt 196 - Parent: Reviewing Family Values with Purpose

Actor: You are a leadership consultant who helps families connect their values to everyday decisions.

Background: I want to make sure our routines and rules truly reflect what we value as a family.

Call to Action: Create a quarterly reflection exercise that helps us evaluate whether our daily actions match our stated family principles, and suggest one way to adjust when they do not align.

Data Check: If you need more information, ask me questions before proceeding.

Prompt 197 - Parent: Practicing Empathetic Accountability

Actor: You are a family communication coach who teaches gentle, reflective accountability techniques.

Background: I want to hold myself and my children accountable without creating guilt or defensiveness.

Call to Action: Provide a script for leading a family reflection after a challenging week that focuses on learning, responsibility, and encouragement instead of criticism.

Data Check: If you need more information, ask me questions before proceeding.

Prompt 198 - Parent: Developing a Growth-Oriented Family Routine

Actor: You are a personal development specialist who helps parents design habits that encourage continuous growth.

Background: I want to make reflection a natural part of our daily and weekly routines so it feels effortless.

Call to Action: Suggest a morning or evening routine including a brief personal reflection, one family gratitude statement, and a goal for the next day.

Data Check: If you need more information, ask me questions before proceeding.

CHAPTER 31

Leading Forward with AI

L eadership in your home is not something you master and move on from. It is a living process, always unfolding as your children grow and as life reshapes what your family needs from you. Reflection helps you see where you have been and where you want to go next. Living the lessons daily requires steady, practical, and adaptable support. Artificial intelligence serves as a reliable tool, to help you stay intentional even when your energy, memory, or focus waver.

AI is not the voice of authority. It does not replace intuition, faith, or family wisdom. It is simply a structured system designed to make the invisible visible, to hold reminders, patterns, and possibilities in front of you when daily life gets noisy. It can help you organize what you already know, track your progress, and remind you to return to your values. In other words, AI does not lead your family. It helps you lead with greater clarity and calm.

Leadership at home begins with emotional intelligence before action. When you use technology with this mindset, you strengthen awareness rather than dependence. For example, you can use AI to help you pause before reacting. You might ask it to help you identify patterns in your stress or to create calming reminders throughout the day. It could store reflective prompts you revisit once a week, or summarize themes from your journal entries so you notice what has been weighing on your heart. A reliable tool does not tell you what to feel; it helps you make sense of what you already feel so you can respond thoughtfully instead of impulsively.

The emotional tone you bring into your home is also supported by structure. A consistent tool can quietly reinforce the atmosphere you want to cultivate. You might schedule gentle notifications like "Take a breath before walking in the door" or "Notice something you appreciate about your child today." These reminders seem simple, but they shape energy. They help you remember calm is not a personality trait, it is a habit of awareness. AI becomes the background rhythm that steadies your tone, keeping your leadership intentional even on difficult days.

Expectations and boundaries give families a sense of safety and rhythm. They also keep technology in its proper place. When you use AI intentionally, you define its role clearly: it serves, it organizes, and it tracks. It does not dictate. For instance, you can use it to store family routines, create shared schedules, or break big goals into manageable steps. It might help you list bedtime tasks, track school projects, or plan family meetings. Boundaries in technology use remind everyone that leadership still comes from you. AI is the filing cabinet, not the compass. You decide the direction.

Empathetic accountability grows when you use AI to stay consistent with what you value. A reliable tool can help you remember to follow up on important moments. You can set reminders to revisit a difficult conversation, to apologize when needed, or to celebrate small wins. These reminders protect your relationships from being swallowed by busyness. They also help you lead with integrity. Leadership is rarely about grand gestures, it is about following through when no one else remembers. When technology helps you do that, it strengthens the trust between your words and your actions.

A growth mindset leader sees tools as opportunities for improvement, not replacements for effort. AI gives you the ability to transform your reflections into forward movement. You might use it to brainstorm ideas for connecting with your teen, to outline new traditions, or to explore activities that fit your child's interests. You can ask it for ways to explain a rule more clearly or to help you craft a message that blends firmness with empathy. The tool responds instantly, but the wisdom still belongs to you. The goal is not efficiency for its own sake, it is consistency rooted in care.

A practical way to use AI as a reliable tool is through simple daily or weekly interactions. You can store a family reflection journal and review it each month to see how your tone, routines, and leadership choices have evolved. You can record family goals and ask the system to remind you to check progress every Sunday evening. You can even create conversation scripts for common moments, such as explaining a boundary, giving feedback, or helping your child recover from disappointment. When these tools become part of your rhythm, reflection stops being a task and starts being a lifestyle.

Imagine saying, "Remind me every first Sunday to do our family reflection check-in." Or, "Help me phrase an apology that feels sincere but balanced." Or, "Suggest gentle ways to explain why we are changing bedtime routines." None of these requests give control away. They simply make leadership lighter. They create room for you to show up as the calm, present parent you intend to be, especially on days when your patience or creativity feel worn thin.

As you continue leading your family, remember AI is not a voice of judgment or instruction. It is a reliable notebook that listens and reflects back what you have taught it. It can hold your questions, track your ideas, and help you translate reflection into action. It does not replace your judgment. It strengthens it.

In many ways, AI mirrors the leadership principles you have practiced throughout this book. It helps you stay emotionally aware, maintain a steady tone, uphold consistent boundaries, hold yourself accountable with grace, and keep growing even when life demands more than you planned to give. The most meaningful leaders never stop learning, and the most intentional families never stop refining how they love. AI simply makes learning visible.

Used wisely, this tool keeps you grounded in your humanity while expanding your capacity. It holds your reminders, your progress, and your promises until you are ready to act on them. When you treat reflection as both a mindset and a method, supported by a reliable tool that never tires, you become the kind of leader every family needs, a leader who keeps growing in wisdom, grace, and intention.

Prompt 199 - Parent: Organizing Reflections Over Time

Actor: You are a family leadership coach who helps parents organize their reflections and insights.

Background: I have been journaling about my parenting experiences, and I want to see patterns and growth over time.

Call to Action: Help me create a simple digital reflection log that organizes my monthly notes by theme, such as emotional tone, boundaries, and connection. Include suggestions for labeling or tagging entries for easy review later.

Data Check: If you need more information, ask me questions before proceeding.

Prompt 200 - Parent: Scheduling Reflection Reminders

Actor: You are a time management specialist who helps parents stay consistent with reflection habits.

Background: I want to make family reflection a habit but often forget to pause for it.

Call to Action: Create a reminder system that prompts me weekly and monthly to review our progress and revisit family goals. Include gentle message examples for the reminders needed to inspire reflection rather than pressure.

Data Check: If you need more information, ask me questions before proceeding.

Prompt 201 - Parent: Planning Calm Conversations

Actor: You are a family communication trainer who helps parents prepare for calm, empathetic conversations.

Background: I want to plan how to discuss changes or challenges with my children without losing patience or authority.

Call to Action: Help me draft three example conversation outlines that combine clarity and warmth when discussing new expectations or resolving tension at home.

Data Check: If you need more information, ask me questions before proceeding.

Prompt 202 - Parent: Tracking Family Goals and Progress

Actor: You are a leadership strategist who helps families translate big goals into achievable steps.

Background: Our family sets intentions but sometimes loses track of progress because we do not have a system to measure it.

Call to Action: Design a family progress tracker that breaks big goals into smaller weekly actions and includes a reflection question for each week.

Data Check: If you need more information, ask me questions before proceeding.

Prompt 203 - Parent: Encouraging Accountability with Kindness

Actor: You are a parenting mentor who helps families maintain accountability through gentle follow-up.

Background: I want to remember to follow up after big conversations or moments of discipline to show my child that growth takes time.

Call to Action: Create a checklist of reflective follow-ups I can schedule, such as gratitude notes, repair conversations, or appreciation moments to strengthen empathy and consistency in our home.

Data Check: If you need more information, ask me questions before proceeding.

Prompt 204 - Parent: Summarizing Monthly Family Themes

Actor: You are an organizational coach who helps parents translate reflection data into insight.

Background: I keep notes about my family's routines, moods, and progress, but I have trouble identifying patterns.

Call to Action: Review my journal entries or notes and generate a short summary of the key themes or changes that emerged this month. Provide one growth insight and one actionable step for next month.

Data Check: If you need more information, ask me questions before proceeding.

CHAPTER 32

When the Story Doesn't Match the Dream

I am including this chapter to provide a balanced view, because no one, absolutely no one, is a perfect parent. At first glance, this chapter may not seem like it belongs in a book about AI powered leadership for families. But it does. It belongs because it reflects a sad reality for many parents and grandparents, one that often unfolds quietly behind closed doors and in unspoken pain. Even the most dedicated and loving parents can find themselves facing estrangement, wondering how a bond that once felt unbreakable could grow silent.

This chapter is not written to frighten you but to prepare and inspire you. Parenting is not a guarantee of loyalty or closeness. You can give everything you have, your love, your time, and your sacrifices, and still face distance or misunderstanding. A truth that does not make you a failure. It makes you human. I want to equip you with wisdom, not only for the joyful seasons of parenting but also for the seasons that test your heart. If this ever becomes part of your story, I want you to have the tools, language, and courage to find peace and begin again. And if this is not your story, share it with a friend, neighbor, or colleague who may be experiencing family estrangement. Your empathy and understanding could become the reminder they need that hope and healing are still possible.

Even the most devoted parent can find themselves facing distance they never expected. You can pour everything you have into your children, giving your time, your wisdom, and your best intentions, and still watch them choose separation. Doing your best has never guaranteed a particular outcome, because parenting is not a formula; it is a commitment. We do it not for applause or assurance, but because it is part of who we are. Parenting is leadership in its purest form, a daily decision to show up, love well, and guide with integrity even when the return is uncertain. The goal is not control, but character. You must parent faithfully, not perfectly, but trusting love, when given sincerely, always leaves a mark.

Every parent begins the journey with a dream. We imagine a home filled with laughter, traditions lasting for generations, and children who grow into adults who still want to come back and sit at our tables. We picture family as an unbreakable circle; but life has a way of challenging the stories we write in our hearts. Some parents give everything they have, love, structure, sacrifice, and wisdom, and still find themselves facing silence from the very children they once tucked into bed.

We are living in a time when family estrangement has become almost fashionable. Social media rewards detachment, with endless voices telling people to cut off anyone who makes them uncomfortable, even those who once cared for them. The phrases "protect your peace" and "go no contact" are now worn like badges of honor. But cutting off a family without sincerely trying to heal is not empowerment; it is a reason to grieve. When children disown parents and parents disown children, something sacred has been lost. These are not signs of growth but of pain left unaddressed. When we choose disconnection over repair, it can cause sepsis of the heart, slowly poisoning the very love we were meant to protect.

What is often forgotten (or never even considered) is estrangement does not just separate individuals; it fractures generations. When parents and children stop speaking, grandchildren lose access to grandparents who hold history, wisdom, and unconditional love. It robs the young of the chance to experience a kind of affection that is slower, gentler, and often freer than what parents can give. Being a grandparent is not the same as being a parent. Grandparents see the family story from a wider lens. They love without the daily weight of discipline and offer a sense of continuity that connects the past, present, and future.

I experienced this lesson personally. There was a time when my mother-in-law and I could not seem to find common ground. We disagreed often and saw the world through very different lenses. Yet, I made the decision to let her relationship with my children stand on its own. I did not interfere, because I respected what she could bring to their lives. She invited them to her home to spend time with her, creating their own rhythm and connection. I never prevented her connection with my children because it would have been petty and selfish.

Just to be clear, there are situations of abuse that justify oversight or separation. Approximately, ten percent of parent-child fractures stem from physical, emotional, or sexual abuse. In those cases, the healthiest and safest choice is for the parent to remain distant. Abuse and maltreatment can cause lasting estrangement and even lead to foster care placement, especially when the harm is severe or repeated.

However, this is not the type of estrangement I am addressing in this chapter. I am speaking to families divided by differences in values, beliefs, or evolving life choices, where love still exists, but understanding has faded. If I had my way, I would encourage every family in this situation to work with a skilled therapist and treat family repair as a responsibility, not an option. Because if healing happens apart from one another, is it truly healing, or simply retreating?

If you are a parent, or grandparent who has experienced estrangement, know this: You are not a failure. You are an imperfect, tender, human. The silence does not erase the love you gave, and it does not cancel the years of leadership, care, and sacrifice that shaped your family. It simply means there are wounds that need tending. Healing begins by pausing before reacting, by noticing what you feel without letting those emotions lead you astray. Grief, anger, or regret can be powerful teachers when you sit with them long enough to understand their message. Emotional intelligence is not about controlling what you feel, but about choosing how you respond.

Even in silence, your emotional tone still matters. The energy you carry becomes its own kind of message. Bitterness repels; calm invites. When you do the quiet work of healing yourself, you create emotional safety without speaking a word. Children who once felt distant often sense change long before they acknowledge it. Peace has a presence, and it can become a silent invitation for reconciliation.

Part of healing involves holding steady boundaries with grace. Healthy love neither chases nor withdraws. It remains consistent, grounded, and clear. You can love your child without accepting harmful behavior. You can offer forgiveness without denying your pain. You can keep the door open without standing in it. Boundaries protect what is sacred in you while leaving space for relationships to return when both hearts are ready.

Rebuilding trust requires humility, accountability, and empathy. Healing is not about pretending the past did not happen; it is about facing it with honesty and love. You might write a letter that acknowledges the hurt, not to erase the past, but to show you understand it. You might say, "I can see how you felt, and I wish I had responded differently." That kind of accountability transforms pain into understanding. It also models leadership through vulnerability, showing that true strength comes from the willingness to grow.

Growth is the quiet companion of every healing parent. There will be days when it feels like you are standing still, and others when you see small signs of change. Sometimes it takes years for a child to recognize the depth of what you tried to give. Sometimes they will rewrite the family story in ways that make you ache. It hurts deeply to be misunderstood, but your story is not finished. Growth often happens in the unseen spaces, where patience and faith hold the line until perspective returns.

When the story does not match the script, love finds new forms. If reconciliation has not yet come, you can still pour your wisdom into the world. Mentor a young parent. Volunteer at a school or youth center. Offer support to families who are struggling to hold on to each other. Love is not limited by blood; it grows wherever it is given room to breathe. In addition to my beautiful children and grandchildren by birth, I have also found this truth in my own life through my neighbor's daughter, my fictive granddaughter, who is mine by choice and by love. She reminds me that family can be created in unexpected ways and that connection, when nurtured with care, can always flourish.

Healing from estrangement is slow, deliberate work. It takes guidance, community, and courage. It may mean learning new communication skills, practicing forgiveness, and rediscovering joy in unexpected places. AI tools can support you in this process, helping you prepare for difficult conversations, practice reflection, or draft messages with clarity and compassion. The goal is not perfection but peace, and to communicate from calm rather than reaction.

Leadership in love means continuing to show up for growth, even when no one sees the effort. It means choosing calm over resentment, accountability over blame, and hope over despair. Healing is contagious. When one person begins to grow, the atmosphere begins to shift. Whether reconciliation happens now or in some distant season, your work still matters. The peace you cultivate today may become the bridge your children one day walk across to come home.

Estrangement may be fashionable, but love never goes out of style. It endures through silence. It matures through heartbreak. It waits with grace rather than judgment. The form of your family may change, but your calling as a parent remains the same, to love with wisdom, to forgive with courage, and to keep leading with faith that healing is still possible.

When repair is not possible, life itself can still be. Parents and grandparents deserve to live fully, even in the absence of reconciliation. Healing does not mean waiting for a call that never comes; it means choosing joy in the life you have now. Take grand vacations. Join clubs that make you laugh. Host dinners, dance at concerts, and make new memories with people who celebrate your presence. Live out loud. Every act of joy becomes an act of resistance against bitterness. When you keep living with light in your spirit, you remind yourself and your children, whether they see it or not, that love still wins.

If you are still reflecting, living lovingly, and still learning, then you are still leading. This is the kind of leadership the world needs most, the kind that believes restoration is possible, that healing is holy, and that love will always have a home.

AI Prompts in the ABCD Framework

Prompt 205 - Parent: Writing a Letter of Repair

Actor: You are a family communication coach who helps parents write heartfelt letters that express accountability and care.

Background: I want to reach out to my estranged adult child in a way that acknowledges pain but invites healing.

Call to Action: Help me write a short letter that expresses understanding, avoids blame, and keeps the door open for future connection.

Data Check: If you need more information, ask me questions before proceeding.

Prompt 206 - Parent: Practicing Calm Before Contact

Actor: You are a parenting mentor who helps adults prepare emotionally for difficult family interactions.

Background: I often feel anxious or defensive before reaching out to my child.

Call to Action: Create a short reflection or breathing routine I can use before writing, calling, or responding to my child so that I lead with peace instead of fear.

Data Check: If you need more information, ask me questions before proceeding.

Prompt 207 - Parent: Setting Healthy Emotional Boundaries

Actor: You are a counselor who specializes in family dynamics and emotional resilience.

Background: I want to maintain love for my child without losing my emotional balance. But, my stomach clinches when I see their phone numbers pop up on my phone. I have been invited and uninvited to events, deprioritized, and dismissed too often.

Call to Action: Design a list of emotional boundaries and self-care practices that protect my peace while keeping my heart open for reconciliation.

Data Check: If you need more information, ask me questions before proceeding.

Prompt 208 - Parent: Rehearsing Reconnection Conversations

Actor: You are a dialogue coach who helps families rebuild trust through intentional communication.

Background: I want to prepare for the day when my son and I finally talk again, but I am unsure how to begin. He won't let me send cards or gifts, claiming it's menacing. It feels like someone is in his ear, encouraging estrangement.

Call to Action: Write three sample conversation openers that communicate love, humility, and readiness to listen, along with one reflective question I can ask to keep the conversation balanced.

Data Check: If you need more information, ask me questions before proceeding.

Prompt 209 - Parent: Reframing the Parenting Story

Actor: You are a family therapist who helps parents reframe painful experiences with clarity and compassion.

Background: I am struggling to accept that I gave my best as a parent, even though my child has chosen distance.

Call to Action: Help me reframe my parenting story in a way that honors my effort and growth without blaming myself for my child's choices.

Data Check: If you need more information, ask me questions before proceeding.

Prompt 210 - Parent: Affirmations for Healing Parents

Actor: You are a wellness coach who specializes in affirmations for emotional resilience.

Background: I often feel unappreciated and isolated after becoming estranged from my child.

Call to Action: Suggest daily affirmations that reinforce self-worth, peace, and hope for reconciliation.

Data Check: If you need more information, ask me questions before proceeding.

Prompt 211 - Parent: Evaluating Healthy Boundaries

Actor: You are a parenting consultant who helps adults make balanced decisions about family boundaries.

Background: I am trying to decide whether setting a new boundary with my child is protective or unnecessarily harsh.

Call to Action: Guide me through evaluating this boundary so that it aligns with both my peace and my values.

Data Check: If you need more information, ask me questions before proceeding.

Prompt 212 - Parent: Writing a Letter for Emotional Release

Actor: You are a grief counselor who helps people find closure through reflective writing.

Background: I have strong feelings I want to express to my child, but I may never send the message.

Call to Action: Help me write a letter that releases my emotions, offers forgiveness, and helps me heal privately.

Data Check: If you need more information, ask me questions before proceeding.

Prompt 213 - Parent: Living with Meaning Beyond Estrangement

Actor: You are a life coach who helps parents rediscover purpose after family distance.

Background: I want to live a meaningful life that is not dependent on my child's approval or involvement.

Call to Action: Suggest ways I can build a fulfilling routine, connect with supportive people, and find joy through personal growth and service.

Data Check: If you need more information, ask me questions before proceeding.

Prompt 214 - Parent: Finding Fulfillment as an Empty Nester

Actor: You are a purpose coach who guides adults through major life transitions.

Background: I am an empty nester experiencing grief instead of relief, and I want to find renewed meaning.

Call to Action: Suggest fulfilling activities, community roles, or long-term goals that bring joy, contribution, and purpose to this new stage of life.

Data Check: If you need more information, ask me questions before proceeding.

Prompt 215 - Parent: Extending Parenting Beyond My Household

Actor: You are a mentorship strategist who helps parents share their wisdom with others.

Background: I want to use my parenting experience to support young people or families outside my home.

Call to Action: Suggest ways I can apply my leadership, love, and skills in mentoring, teaching, or community service.

Data Check: If you need more information, ask me questions before proceeding.

Prompt 216 - Parent: Volunteering to Nurture Others

Actor: You are a volunteer coordinator who connects adults with opportunities for service.

Background: I have love and energy to give but no immediate family to focus it on.

Call to Action: Suggest meaningful volunteer opportunities where I can nurture others' growth, such as mentoring teens, tutoring, or supporting caregivers.

Data Check: If you need more information, ask me questions before proceeding.

Prompt 217 - Parent: Parenting and Imposter Syndrome

Actor: You are a family leadership coach who helps parents overcome self-doubt and strengthen confidence in their parenting identity.

Background: I often feel like I am failing as a parent, even though I try my best. When things go wrong with my child, I question whether I am doing enough or if I even deserve to be called a parent. I want to stop comparing myself to others and start leading my family with calm confidence instead of guilt or fear.

Call to Action: Help me identify the signs of imposter syndrome in my parenting and create daily affirmations, reflection questions, and practical steps to rebuild confidence in my role as a parent leader. Include one exercise I can do each week to strengthen self-trust.

Data Check: If you need more information, ask me questions before proceeding.

The Beginning of Your Next Chapter

If you have made it this far, take a deep breath and smile. You did it. You have just completed *AI Powered Leadership for Modern Families,* and I hope you feel what I feel every time I reach a parenting milestone—proud, tired, inspired, and maybe a little amused that we ever thought this job would be easy.

When I began writing this book, I did not want to hand you a rulebook filled with perfect-parent commandments. There are already too many of those. My goal was to give you a toolkit, a set of flexible, practical strategies and AI prompts that grow with you, your children, your grandchildren, and the ever-changing world around you.

Inside these pages, you have explored over two hundred prompts designed to help you lead your family with empathy, clarity, and confidence. Here is the real secret: The two hundred prompts found in this book were never the finish line. They were the beginning. The moment you start shaping these ideas to fit your family's rhythm, your values, and your vision, you graduate from following prompts to crafting your own. That is where the transformation begins. Every family has its own stories, traditions, and challenges. What works beautifully for mine may not work for yours, and sometimes not even for me. The beauty of AI is it lets you customize. It is your round-the-clock research mechanism, creative sounding board, and planning assistant all in one.

Before you rush back into the rhythm of daily life, take a moment to hold onto what you have learned. The insights, tools, and reflections you have gathered are not meant to fade when the last page closes. They are seeds of wisdom that will keep growing as you do. Parenting will continue to surprise, stretch, and sometimes overwhelm you, but the lessons you have practiced about empathy, clarity, and calm leadership are your steady foundation. Keep them close as you move forward and remember that every new challenge is another chance to lead with love and begin again.

Prompt 218 - Parent: Leadership Reflection

Actor: You are a Family Leadership Coach and Reflective Partner.

Background: I have finished reading *AI Powered Leadership for Modern Families.* I have learned practical strategies, emotional leadership principles, and how to use AI to guide thoughtful parenting. I want to carry these lessons forward as I continue growing as a parent and leader in my family.

Call to Action: Help me reflect on what I have learned from this book and identify three meaningful ways to apply the ABCD Framework in my daily life. I want ideas that help me stay grounded, keep leading with love, and continue using AI as a tool for self-awareness and growth.

Include a brief affirmation I can repeat when parenting feels hard or uncertain.

Data Check: If you need more information, ask me questions before proceeding.

See? You have already got this. The prompts in this book were never meant to end; they were meant to spark a lifelong skill, learning how to ask better questions.

AI does not replace you; it reveals more of what is possible for you. It gives you access to collective knowledge that stretches across psychology, neuroscience, education, communication, and emotional intelligence. It is like having access to the world's largest parenting library, open twenty-four hours a day, with no overdue fees and no judgment. When you use AI as a reference rather than a replacement for your wisdom or compassion, you become calmer and more creative. It helps you pause before reacting, find words when emotions run high, and explore choices you might have overlooked. You are no longer parenting in isolation. You are leading with insight that spans generations, cultures, and disciplines, all available in real time.

The beauty of this moment in history is you can merge timeless parenting wisdom with modern technology. You get to model lifelong learning, showing your children that curiosity and humility are stronger foundations than perfection. You are not just raising children; you are preparing future leaders who understand how to think, not just what to think.

AI is a remarkable tool, but it is not a therapist, doctor, or oracle. It cannot hug your child, hold your hand through loss, or know the sound of your child's laughter. It cannot replace the human connection that makes love feel safe and real. Remember to use discernment and wisdom. Seek counseling, medical advice, or spiritual guidance when needed. Let AI guide you toward clarity, but never let it replace your intuition.

Parenting has never been about perfection. It is about presence, practice, and progress. Some days will look like harmony and laughter. Others will look like cereal for dinner and bedtime negotiations worthy of an Olympic event. Either way, you are doing the work of leadership, the steady, loving, and sometimes messy kind that shapes character, trust, and belonging.

You already have everything you need to lead your family forward. Use the tools. Keep asking thoughtful questions. Keep showing up with grace, courage, and curiosity. This may be the final chapter of the book, but it is only the beginning of your next chapter as a parent leader. Because leadership at home is never about knowing it all. It is about staying teachable, staying human, and choosing to love again tomorrow, no matter how today went.

A Personal Note from the Author

If you are reading this, I hope you understand what an act of love it is just to stop and reflect. Most people parent on autopilot, reacting to the chaos of the day like short-order cooks in the busiest kitchen on Earth. But you paused long enough to think. You cared enough to ask hard questions and to keep learning. That pause is not small. That pause is leadership.

There are no diplomas, no final exams, and no graduation day in parenting. No one drapes a sash over your shoulders and declares, "You've mastered the raising of humans." Parenting is an infinite project made of experiments, trial runs, and small miracles disguised as ordinary days. It is humbling, unpredictable, sometimes thankless, and always unfinished.

The hardest truth of all is that parenting continues even when your children pull away. It continues when they ignore your texts, roll their eyes, or forget your birthday. It continues when you are not invited to milestones you once dreamed of sharing. Love does not end when it is unreturned. It becomes quieter. It moves underground and grows roots through prayer, patience, and faith that one day, the bridge might be rebuilt. Leadership at home means loving your children even when you must do it from a distance.

If I could leave you with one message, it would be this: **Keep loving and keep trying.**

Even when you slip back into old patterns. Even when your kids groan at your new routines or disappear when you announce a "family meeting about feelings." Even when it feels like you are the only one doing the work, and the progress is invisible, keep trying. Because that is what real leadership looks like, steady, imperfect effort. Every time you choose to try again, you teach your children what resilience looks like. They may not say it now, and they certainly will not post it online, but they are watching. They are learning how to recover from setbacks, how to persist, and how to keep showing up with love instead of bitterness. Those lessons will last long after childhood ends.

Keep loving, especially when it feels impossible. Love when it is awkward. Love when it is quiet. Love when you are running on empty and when you feel unseen. The love you give today is not a transaction; it is a seed. It may take years to bloom, but its roots will hold firm. One day, when your child faces their own heartbreak or confusion, that same love will rise in them as strength.

You are not chasing perfection; you are building a legacy. Every time you show up, every time you forgive yourself and try again, you are shaping the emotional inheritance your children will carry into their own families. True leadership is built on consistency, compassion, and courage, one imperfect moment at a time.

Your kids will not remember every lecture or perfectly executed rule. They will remember your tone. They will remember how you tried to listen. They will remember that you said, "Let's try again." Those moments become their compass when life gets complicated.

If you start to feel alone, remember this: Every parent who picked up this book is walking the same road. We are all trying to love better, to lead better, to grow alongside the people who matter most. There is no shame in stumbling. The wisdom comes in standing back up.

Parenting can feel isolating because the world shows only the highlight reels. No one posts the meltdowns, the guilt, or the nights spent wondering, "Am I doing this right?" Behind every polished photo is another parent holding it together the best they can. Even the calm, organized, seemingly unshakable parents you admire have moments of doubt and overwhelm. You are not behind. You are in progress.

Progress is the only measure that matters. The fact you are reading these words means you are still 'in the work'. You are still showing up, still reflecting, still growing. That is leadership in motion. The best leaders are not defined by how few times they fall but by how faithfully they rise with purpose.

When you doubt yourself, remember this: the struggle is proof that you care. It means your heart is still in it. You are not failing; you are refining. You are learning how to love in real time, under pressure, and often without applause.

Every parent you know is trying to do the same thing. We are all learners in the same class, just at different desks. You belong to a worldwide community of imperfect, exhausted, hopeful humans who refuse to give up on their children or themselves. That is not failure. That is family.

So, keep this book close, not as a rulebook, but as a mirror and a companion. Let it remind you progress is not linear, reflection is brave, and growth is a lifelong practice. Return to these pages whenever you need encouragement. Laugh at the parts that sound too familiar. Take comfort in knowing no one has it all figured out. And, when you are ready, join me again for the next conversation. We will keep learning together, one prompt, one pause, one ordinary, beautiful day at a time.

You do not have to be perfect. You only have to be present.

Keep the faith. Keep showing up.

That is leadership, and that is love.

Final Prompt: Because I Care

Actor: You are a Family Leadership Coach and Creative Partner.

Background: I have developed the ABCD Framework as both a philosophy and a tool to help parents lead their families with clarity, empathy, and purpose. It is the culmination of lessons learned through trial, triumph, and lots of reflection. I want to offer it to parents as a true gift, something that inspires hope, encourages grace, and strengthens their ability to lead with love.

Call to Action: Help me discover how to deliver this gift in a way that feels uplifting and personal. I want parents to feel seen, supported, and empowered to lead their families in a positive direction, and I want their children to experience the ripple effect of that growth.

Data Check: If you need more information, ask me questions before proceeding.

References

Bloomfield, L., & Barnhardt, R. (1985). Let's read. Addison-Wesley.

Druga, S., Bickmore, T., Hiniker, A., Vu, S., Likhith, E., & Qiu, T. (2022). Family as a third space for AI literacies: How do children and parents learn about AI together? University of Washington.

Freeman, J. B. (2006). *Taking charge of your positive direction.* Trafford Publishing.

Escoredo, M. C., Mostovoy, K., Schickler, R., Bechtel, A., Shagan, J., & Bunge, E. L. (2025). Enhancing parental skills through artificial intelligence-based conversation agents. *Family Relations, 74*(3), 1250-1265.

O'Dell, S. (1960). Island of the blue dolphins. Houghton Mifflin.

Szondy, M. B., & Magyary, A. (2025). Artificial intelligence in the family system: Possible positive and detrimental effects on parenting, communication and family dynamics. *European Journal of Mental Health, 20.*

www.ingramcontent.com/pod-product-compliance
Lightning Source LLC
Chambersburg PA
CBHW071707120626
46550CB00001B/136